MATHEW BOUDREAUX AND TARA J. CURTIS

FABRIC WEAVING

play with color & pattern

12 PROJECTS · 12 DESIGNS TO MIX & MATCH

stashBOOKS®

an imprint of C&T Publishing

Text and photography copyright © 2022 by Mathew Boudreaux and Tara J. Curtis

Photography and artwork copyright © 2022 by C&T Publishing, Inc.

Publisher: Amy Barrett-Daffin

Creative Director: Gailen Runge

Senior Editor: Roxane Cerda

Editor: Kathryn Patterson

Technical Editor: Julie Waldman

Cover/Book Designer: April Mostek

Production Coordinator: Zinnia Heinzmann

Production Editor: Jennifer Warren

Illustrator: Kirstie Pettersen

Photography Coordinator: Lauren Herberg

Photography Assistant: Gabriel Martinez

Cover photography by Mathew Boudreaux

Instructional photography by Mathew Boudreaux and Tara J. Curtis; lifestyle and subjects photography by Lauren Herberg of C&T Publishing, Inc., unless otherwise noted

Published by Stash Books, an imprint of C&T Publishing, Inc., P.O. Box 1456, Lafayette, CA 94549

Library of Congress Cataloging-in-Publication Data

Names: Boudreaux, Mathew, 1976- author. | Curtis, Tara J., 1976- author.
Title: Fabric weaving : play with color & pattern; 12 projects, 12 designs to mix & match / Mathew Boudreaux and Tara J. Curtis.
Description: Lafayette : Stash Books, [2022] | Summary: "Includes how to weave simple classic patterns as well as more complex triaxial designs and learn how to incorporate one-of-a-kind woven textiles into any project or choose from one of several projects in the book"-- Provided by publisher.
Identifiers: LCCN 2021062348 | ISBN 9781644031681 (trade paperback) | ISBN 9781644031698 (ebook)
Subjects: LCSH: Weaving. | Weaving--Patterns.
Classification: LCC TS1490 .B73 2022 | DDC 677/.028242--dc23/eng/20220228
LC record available at https://lccn.loc.gov/2021062348

Printed in the USA

10 9 8 7 6 5 4 3 2

dedication

This book is dedicated to all our weaving friends throughout the years. Thanks for sharing your skills and creativity with us.

acknowledgments

Thank you, Jennifer Hoskisson, for introducing us to one another on Instagram so many years ago. You shoulder all the blame and receive all the credit! We would not be good friends today if it weren't for you.

Speaking of Instagram, a special thank you is due to all the folks who have commented on our posts of woven makes with the same handful of questions: "How do you make your strips? What is that tool you are using to weave? How does it stay together? What will you do with it? Do you quilt it when you're done?" We keep coming up with ways to provide you with answers without having to repeatedly type into a social networking application: tutorials, magazine projects, podcast interviews, patterns, videos … and now we've written a book! You just keep pushing us to find more ways to provide you with answers, and we grow and meet new challenges along the way as a result. Thank you.

CONTENTS

BIAXIAL WEAVING   20

TRIAXIAL WEAVING   72

Small Houndstooth Cross-Body Bag 48

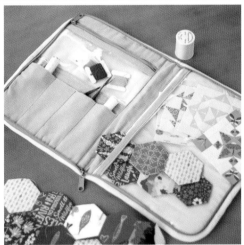

Large Houndstooth Zipped Case 56

Silly Silo Hoop Art 68

Weave Into Denim Jacket Back Panel 98

Woven Stars Messenger Bag 106

Gradient Diamonds Woven Clutch 114

INTRODUCTION TO FABRIC WEAVING

Fabric weaving appeals to the "do-it-yourselfer," the sewist who is drawn to creating her or his own textiles. It is a way of building your own cut of fabric. Fabric weaving also appeals to the fabric lover, the sewist who lives for the touch and feel of fabric just as much as the color and prints. Throughout the weaving process, you are working the fabric with your hands, and the tactile dimension weaving adds to a project is truly special to experience. Woven pieces are as much fun to hold as they are to display!

A woven panel can be used like any other cut of fabric and worked into almost every kind of pattern. Fabric weaving looks great as a bag accent, couch pillow, or dress pocket. You can make wall art, table mats, storage baskets, or travel cases out of your weaving.

types of fabric weaving

In traditional weaving, the warp is the fiber that is strung onto the loom and held with tension. The weft is the fiber that is woven over and under the warp to create cloth. For the purposes of this book, we have divided weaving up into two types of weaving: biaxial and triaxial.

Biaxial means "two axes" and is created using two layers of strips woven into one another. The weft can be woven into the warp at either a 90- or 45-degree (90° or 45°) angle. Biaxial weaving includes designs as basic as basket weave, and as complex as houndstooth. These weaves often need securing with quilting in order to stay together.

Triaxial means "three axes" and is achieved by weaving two weft layers into the warp for a total of three layers. These weft layers are woven in at a 30 degree (30°) angle. A triaxial weave is the most secure weave and requires less quilting to keep the strips in place.

basic tools

BIAS TAPE MAKER OR SASHER

A tool for making bias tape can be helpful but is optional. It's possible to use your fingers to fold the strip while pressing. Bias tape makers can be fiddly and troublesome, but once you get the hang of them, they save time and fingertips. The Dressmaker Sasher by Pauline's Quilters World is intuitive, and the sharp edges of the acrylic give each fabric strip an extra crisp edge.

IRON AND STARCH

An iron in good working order is a necessity for preparing your strips for weaving. You may also use a bit of light starch, such as Mary Ellen's Best Press, to help with this task.

FOAM CORE BOARD

Elmer's has ½″ foam core available in 20″ × 30″ precut sheets. You will want ½″ thick because you will be holding your strips in place with sharp pins, inserted into the board. Though you will insert them at an angle, the thickness of the board can help ensure that pins won't be poking out of the back of the board and into a tabletop, or worse, your lap!

LIGHTWEIGHT WOVEN FUSIBLE INTERFACING

Fusible interfacing is required and will be used to secure your woven panel. Using a lightweight product will help mitigate bulk. You will want to use woven interfacing as opposed to non-woven because it will move, look, and feel like fabric. Pellon makes a nice lightweight woven fusible interfacing called Shape-Flex 101 (SF101).

PINS

You'll need to use lots of pins, and many of them will get bent. Keep a separate set of pins just for weaving (the bent pins can be reused for weaving). Make sure your pins have a nice sized head on them, because you will be pushing them in and removing them multiple times before, during, and after weaving. Dritz Quilting Pins size 28 work well for weaving. They come in a large box of 500, have a nice sized head on them, and are relatively inexpensive.

WEFTY NEEDLE

The WEFTY Needle is the only tool on the market designed specifically for weaving with fabric strips. The size, shape, and design are all intended to help the weaver produce tighter weaves while reducing wear and tear on the hands. Its tapered end was made specifically for gliding through the multiple layers of raw edges in a triaxial weave. While you can make do with something else, all our weaving is done using a WEFTY Needle.

THAT PURPLE THANG

A guide needle to hold up strips while weaving is very helpful. While That Purple Thang is the perfect tool for this task, you could also try a butter knife, awl, or skewer

PAINTER'S TAPE

We will also be using painter's tape to secure our weaves. We use painter's tape as opposed to masking tape because it stays put but with less chance of leaving adhesive residue.

RULER, MARKERS, AND SCISSORS

A basic acrylic ruler, a dark colored marker, a heat erasable pen, and good fabric scissors are indispensable in fabric weaving projects.

fabric selection

When selecting fabric for weaving, the preference is for substrates that are lightweight and tightly woven. We prefer something lightweight because a woven panel can become heavy with all the layers. For example, when creating a triaxial weave, you will be working with six layers of fabric and a layer of interfacing! Some loose threads are inevitable when weaving raw edged strips together. A loose weave, such as homespun fabric, can fray considerably, adding frustration to the process. Quilting cottons, shot cottons, and batiks lend themselves very well for weaving.

Solids, blenders, and prints that read as solid all work very well for weaving. Small scale prints are best, as the surface area of the fabric showing is very small. The design in your weave will be easiest to appreciate if the fabrics contrast nicely. Keep this in mind when looking for gradients—you will want your light, medium, and dark to be obvious.

PREWASHING FABRIC

You may be a prewasher of fabric, in which case you've already washed, dried, pressed, and put away your fabric anyway. You can go ahead and skip to the next section on quilting.

For those who are unsure, or are anti-prewashing, let's discuss when it's helpful and when it isn't. For projects that will not be quilted, and will be washed eventually, prewashing is a must. You don't want to work hard on prepping and weaving strips only to have shrinkage or bleeding dye ruin the look of your piece after the first wash. If you are mixing substrates (for example mixing denim and quilting cotton), you will also want to prewash. If you are planning to spot clean your finished item (such as a purse or wall hanging) or if you plan to densely quilt it, prewashing may not be necessary.

to quilt or not, that is the question

Quilting is not always necessary to keep strips in place after weaving. When Tara created pockets for a dress using small triaxial weaves in ½″ strips, for instance, the woven parts stayed together after washing with no quilting at all. However, if the weave you are creating has fewer layers, or the strips could easily catch on something while in use, you will want to add some quilting. We have provided a couple of quilted projects in this book for inspiration.

binding

You may use these general instructions for most of the projects in this book that require binding.

1. Join binding fabric strips, right sides together at a 45° angle. Trim the excess at ¼″. Press seam open and trim dog ear triangles. *fig A*

2. Fold in half lengthwise with wrong sides together and press.

3. Lay your project down, right side up. Align the raw edge of one end of your binding strip against the middle of a raw edge of one side of your project. Making sure to backstitch, begin sewing 4 inches from the end of the binding strip (so that there is an unattached "tail" of binding that is at least four inches long). Stop sewing ¼″ from the corner and backstitch. *fig B*

4. Fold the binding strip up at a 45° angle and finger-press. *fig C*

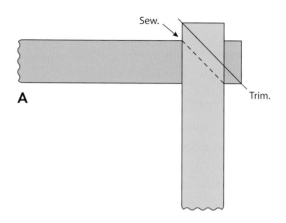

A

Sew.

Trim.

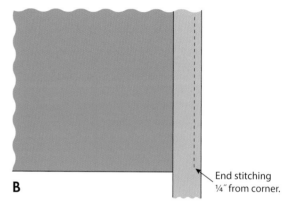

B

End stitching ¼″ from corner.

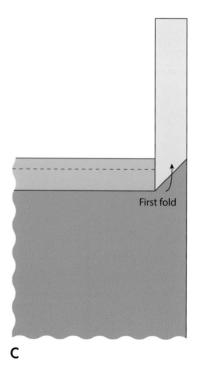

C

First fold

5. Fold the binding strip back down, aligning the raw edges against the next side of your project. Sew from the corner, making sure to backstitch. *fig D*

6. Sew around the next three corners in the same manner. Stop sewing 8 inches from your start point. This should leave another "tail" at the end of your binding.

7. Fold each tail back so the folds touch and press the folds. *fig E*

8. Unfold the binding tails and mark each fold line with a fabric pen or chalk. Place unfolded binding tails right sides together, lining up the marked fold lines. Sew along the line. Do not trim yet.

Finger-press the seam open and re-fold the binding strips. If the binding lays nicely on your project proceed to the next step. If it seems too loose or too tight, rip out the seam and adjust as necessary.

9. When the binding looks good, unfold the binding strip and trim the seam. Press the seam open with an iron. Refold the binding and press. Finish sewing binding onto your project. *fig F*

10. Roll the binding around to the back of your project. Pressing with an iron can make for a professional finish. Use clips or pins to secure the binding, then whipstitch the binding fold to the back of the project.

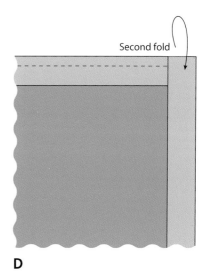

Second fold

D

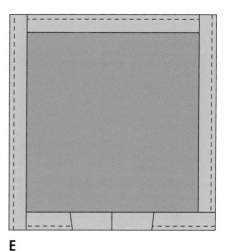

E

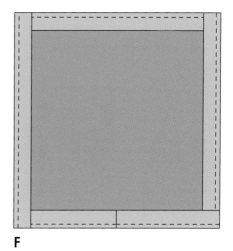

F

WEAVING 101: LEARNING THE BASICS

Let's learn the weaving basics from start to finish with a quick but gorgeous trivet. Your hot pad will feature a simple basket weave in 2 colors or gradients. You'll learn all about prepping strips, drawing a grid, and weaving on a board. Once you're done weaving, you'll secure the weave and cut it into the size and shape you want for your finished project. After this trivet, you'll have all the building blocks to tackle more complicated weave designs!

HAWT PAD

FINISHED SIZE:
10″ × 11″

WOVEN HOT PAD BY TARA CURTIS/WEFTY, MADE USING TWO ANALOGOUS SHOT COTTONS

materials and supplies

MATERIALS

Fabric for weaving: ¼ yard each of 2 colors or gradients

Lightweight woven interfacing: ⅓ yard

Heat-resistant batting such as Insulfleece: ⅓ yard

Backing fabric: 11″ × 12″

Fabric for bias binding: 14″ × 14″

SUPPLIES

Basic tools (page 6)

1″ Sasher or bias tape maker

½″ thick foam core board

1″ WEFTY Needle

cutting

WEAVE FABRICS

• Cut 4 strips 2″ × width of fabric from each fabric (Fabric 1 and Fabric 2).

LIGHTWEIGHT WOVEN INTERFACING

• Cut 1 rectangle 11″ × 12″.

HEAT-RESISTANT BATTING

• Cut 2 rectangles 11″ × 12″.

BIAS BINDING FABRIC

• Prepare bias binding strips by folding the 14″ square diagonally and pressing. Align the 1″ mark of your ruler along the fold and cut. This strip when unfolded will measure 2″. Cut 2″ strips along the diagonal lines of the entire piece.

preparing the strips

1. Lay a Fabric 1 strip wrong side up on the ironing board. Spray lightly with light starch. At one end of the strip, fold the raw edges to the center and press lengthwise.

2. Thread this end into the Sasher and pin the strip to the board. Use a hot dry iron to press the strip while pushing the Sasher down the strip and use your other hand to help feed the strip into the Sasher. An optional step to get extra crisp strips is to lay them so the folded edges are facing down and press them once more. Repeat this until you have 4 beautiful strips ready for weaving from each color. Separate these out into Fabric 1 and Fabric 2 and keep them organized.

preparing the board

1. Using a ruler and dark colored marker, draw a 10″ × 12″ rectangle on the foam core board. Fill this in with a 1″ grid by drawing lines every 1″ horizontally and vertically.

2. Lay the lightweight woven fusible interfacing *adhesive side up* over the grid. You should feel the bumpy side on the top! This is important because you don't want the bumpy side sticking to the board later. You want it to stick to the underside of the *woven strips* later. Pin each corner, inserting the pins at an extreme angle.

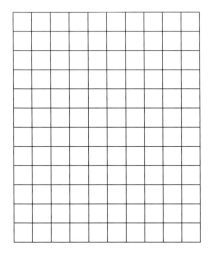

weaving

The basket weave pattern you are creating is made up of 2 layers. The first layer of your woven panel is made up of Fabric 1.

FIRST LAYER

Lay a Fabric 1 strip down on the board, raw edges facing down. You should be able to see the gridlines through the interfacing. Use them as guides to keep the strip straight. Pull the strip taut and pin. Trim. Use the excess for the next strip, until your strip is no longer long enough. Repeat until your grid is covered. Make sure strips are right next to one another without overlapping. You don't want to be able to see the interfacing between the strips. *fig A*

> **Tip: About Pinning**
>
> *Always pin at an extreme angle so that the heads of the pins are flush with the board, for a number of reasons:*
>
> *The pins won't poke out of the back of the board and stick into your lap.*
>
> *The material can't shift on you.*
>
> *The pins don't get in your way while you weave!*

THREADING THE WEFTY

Thread your WEFTY by holding it with the WEFTY logo facing up and inserting a strip into the eye of the needle, folding it back with raw edges together. *fig B*

SECOND LAYER

The second layer of the woven panel is made up of Fabric 2, woven into the first layer at 90°.

1. Thread your WEFTY with a Fabric 2 strip and weave it into one side of the first layer strips. Lift the first layer strips for the WEFTY using the Purple Thang. *fig C*

THE WEAVING SEQUENCE YOU WILL FOLLOW IS UNDER ONE, OVER ONE, REPEATING. Ensure the strip is straight by peeking between the first layer strips to see the gridlines.

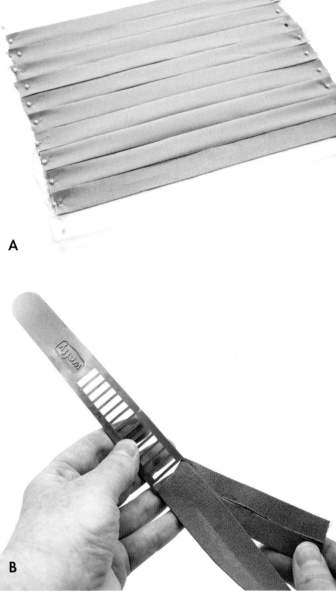

A

B

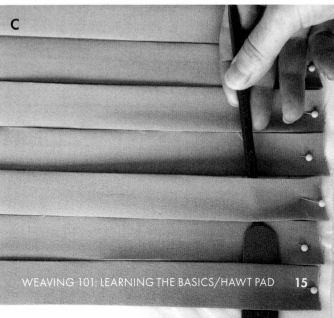

C

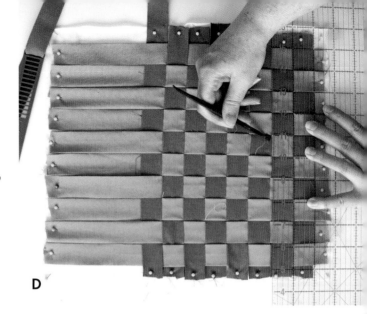

D

2. Pull taut, and then pin at the very end of the strip close to the outside edge of the first layer and trim. Pin just outside of the first layer on the other side of the Fabric 2 strip. Trim this strip just next to the pin.

3. Weave the excess of the Fabric 2 strip into the first layer right next to the strip you just wove in.

USE THE FOLLOWING SEQUENCE: OVER ONE, UNDER ONE, REPEATING.

4. Gripping each side of the strip, gently shimmy the strip back and forth to wiggle it closer to the first strip. Use a ruler and your Purple Thang to finish pushing the strip close to the first, making sure there are no gaps between them, then pin. *fig D*

5. Keep alternating the weaving sequence and straightening the strips until you've finished weaving into the first layer.

6. Carefully trim any loose threads. *fig E*

E

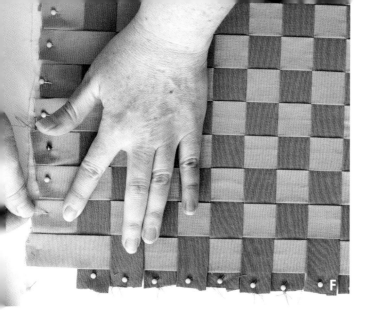

securing the weave

1. Press the panel gently with an iron on the steam setting, being mindful to avoid the pins. This will begin activating the fusible in the interfacing and securing woven strips.

2. Remove the pins by holding the strip down with one finger while pulling out the pin. Make sure all pins are removed! *fig F*

3. Press gently around the edges of the woven panel with the iron on the steam setting. *fig G*

4. Carefully remove the panel from the board by gripping the sides of the interfacing. Move the panel to your sewing machine and baste stitch around all 4 edges using a ⅛″ seam. *fig H*

5. Trim just outside of the basting stitch. *fig I*

If you want to make your trivet hot pad 1″ narrower or shorter, simply baste stitch between the 2 outermost strips on either side and trim just outside of the basting stitch. This will remove the outermost strip while keeping the woven panel together.

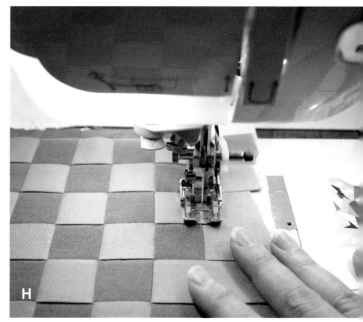

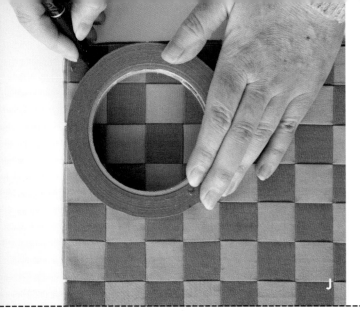

6. Round the corners, using a bowl or roll of masking tape and a heat erasable pen to draw a curve at each corner. *fig J*

7. Baste stitch just inside of this drawn line, then cut just outside of your stitches. *fig K*

K

layering and binding

LAYERING

Lay the backing fabric wrong side up and smooth it out. Center the batting on top. Place the woven panel right side up on top. Trim the excess batting and backing fabric to match the shape of the woven panel. Use clips or basting stitches to hold the 3 layers together. *fig L*

SEWING THE BINDING

Join the ends of the binding strips right sides together (see Binding, page 10) and trim off the dog ears. Lay the binding right-side down on your ironing board, fold in half lengthwise and press to get 1″ single-fold bias binding.

ATTACHING THE BINDING

1. Leaving a 5″ tail, align the raw edge of the binding to the raw edge of the backing side of the trivet hot pad and use clips to secure. *fig M*

The bias binding will easily stretch to bend around the curved corners.

2. Leaving a 5″ tail at the end, trim off the excess bias binding. From the excess, cut a 7″ strip with no seams. Set this aside for the loop.

L

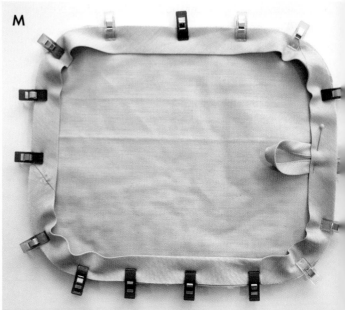

M

MAKING THE LOOP

Unfold, then press the raw edges to the center lengthwise and press. Fold in half again and press. You will now have a ½″ wide double-fold bias strip. Top stitch ⅛″ from the folds to close. Fold this into a loop and secure to a short side of the trivet using a pin. *fig N*

SEWING ON THE BINDING

1. Sew the binding on using a scant ¼″ seam. Press the binding to help train it to wrap around to the front of your trivet. You may need to trim the seam allowance so that the binding will cover the stitches on the woven side. *fig O*

2. Press and clip the binding to the front of the hot pad. Make sure your loop is in the position you want.

3. Attach the binding by sewing a topstitch ⅛″ from the fold.

4. Tara finishes her binding by folding the end of one tail under ¼″, pressing the fold, and then nesting the other tail inside the folded under tail. For a more traditional method, try the binding technique (page 10) to finish your binding. *fig P*

You're done! Go make some mac-n-cheese and use your new Hawt Pad!

N

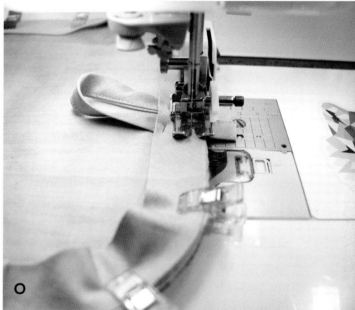
O

P

BIAXIAL WEAVING

DIAMOND TWILL PILLOW

FINISHED SIZE: 19″ × 19″

Once you've learned the basics of weaving with your quilting cottons, look around for inspiration at weavings done using other materials. Bamboo, paper, and yarn are all used to create designs that can be recreated with fabric strips. Parquet flooring, baskets, and chair caning are all sources of design ideas. This hand-woven pillow is inspired by parquet flooring weaves done using strips of bamboo and looks best in two contrasting colors.

DIAMOND TWILL WOVEN PILLOW BY TARA CURTIS/WEFTY, MADE USING CONTRASTING BELLA SOLIDS AND GOLDEN HOUR FABRIC BY RUBY STAR SOCIETY FOR MODA FABRICS

materials and supplies

MATERIALS

Fabric for weaving: ⅔ yard each of 2 contrasting colors or gradients

Lightweight woven interfacing: ⅔ yard

Fusible fleece: ⅔ yard

Fabric for backing and binding: 1 yard

SUPPLIES

Basic tools (page 7)

1″ Sasher or bias tape maker

½″ thick foam core board

1″ WEFTY Needle

Perle cotton (optional)

Milliners needle (optional)

18″ zipper

20″ × 20″ pillow form (or make your own)

2 squares of muslin 22″ × 22″ (if making your own pillow form)

Pillow stuffing (if making your own pillow form)

cutting

WEAVE FABRICS

• Cut 10 strips 2″ × width of fabric from each fabric (Fabric 1 and Fabric 2).

LIGHTWEIGHT WOVEN INTERFACING

• Cut 1 square 20″ × 20″.

FUSIBLE FLEECE

• Cut 1 square 23″ × 23″.

BACKING AND BINDING FABRIC

• Cut 1 square 23″ × 23″.

• Cut 3 strips 2½″ × width of fabric.

preparing the strips

1. Using the instructions from Preparing the Strips (page 14), press 10 beautiful strips ready for weaving from each color.

2. Cut each strip in half so they measure approximately 20˝ long.

preparing the board

1. Using a ruler and dark colored marker, draw a 20˝ × 20˝ square on the on the foam core board. Draw a 2˝ grid on the square.

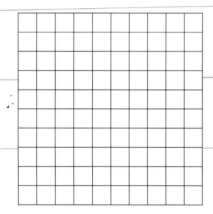

2. Lay the lightweight woven fusible interfacing *adhesive side up* over the grid, following the instructions on Preparing the Board (page 14).

weaving

The weave pattern you are creating is made up of 2 layers. The first layer of your woven panel is made up of Fabric 1.

FIRST LAYER

Lay a Fabric 1 strip down on the board horizontally, raw edges facing down. You should be able to see the gridlines through the interfacing. Use them as guides to keep the strip straight. Pull the strip taut and pin.

Repeat until your grid has 19 strips. Make sure strips are right next to one another without overlapping.

SECOND LAYER

The second layer of the woven panel is made up of 19 Fabric 2 strips, woven into the first layer at 90°.

1. Thread your WEFTY with a Fabric 2 strip and weave it into one side of the first layer strips. Lift the first layer strips for the WEFTY using the Purple Thang.

2. THE WEAVING SEQUENCE DIFFERS FOR EACH STRIP AND IS LISTED BELOW. BEGIN WEAVING STRIP 1 AT THE FAR LEFT OF THE FIRST LAYER.

Strip 1: Under 3, Over 3, Under 3, Over 1, Under 3, Over 3, Under 3

Strip 2: Under 2, Over 3, Under 3, Over 3, Under 3, Over 3, Under 2

Strip 3: Under 1, Over 3, Under 3, Over 5, Under 3, Over 3, Under 1

Strip 4: Over 3, Under 3, Over 3, Under 1, Over 3, Under 3, Over 3

Strip 5: Over 2, Under 3, Over 3, Under 3, Over 3, Under 3, Over 2

Strip 6: Over 1, Under 3, Over 3, Under 5, Over 3, Under 3, Over 1

Strip 7: Under 3, Over 3, Under 3, Over 1, Under 3, Over 3, Under 3

Strip 8: Under 2, Over 3, Under 3, Over 3, Under 3, Over 3, Under 2

Strip 9: Under 1, Over 3, Under 3, Over 5, Under 3, Over 3, Under 1

Strip 10: Over 3, Under 3, Over 3, Under 1, Over 3, Under 3, Over 3

Strip 11: Under 1, Over 3, Under 3, Over 5, Under 3, Over 3, Under 1

Strip 12: Under 2, Over 3, Under 3, Over 3, Under 3, Over 3, Under 2

Strip 13: Under 3, Over 3, Under 3, Over 1, Under 3, Over 3, Under 3

Strip 14: Over 1, Under 3, Over 3, Under 5, Over 3, Under 3, Over 1

Strip 15: Over 2, Under 3, Over 3, Under 3, Over 3, Under 3, Over 2

Strip 16: Over 3, Under 3, Over 3, Under 1, Over 3, Under 3, Over 3

Strip 17: Under 1, Over 3, Under 3, Over 5, Under 3, Over 3, Under 1

Strip 18: Under 2, Over 3, Under 3, Over 3, Under 3, Over 3, Under 2

Strip 19: Under 3, Over 3, Under 3, Over 1, Under 3, Over 3, Under 3

3. Remember to pull each strip taut after weaving and pin in place.

4. After weaving in a strip, grip each side of it and gently shimmy the strip back and forth to wiggle it closer to the strip you wove in before it. Use a ruler and your Purple Thang to finish pushing the strip close to the first, making sure there are no gaps between them, then pin.

5. Carefully trim any loose threads.

securing the weave

1. Press the panel gently with an iron on the steam setting, being mindful to avoid the pins. This will begin activating the fusible in the interfacing and securing woven strips.

2. Place painter's tape around the perimeter, as well as over the Fabric 2 strips. Finger-press firmly in place.

3. Remove the pins by holding the strip down with one finger while pulling out the pin. Make sure all pins are removed! *fig A*

4. Carefully remove the panel from the board by gripping the sides of interfacing. Move the panel to your sewing machine and baste stitch around all 4 edges ⅛″ from the tape. *fig B*

5. Trim just outside of the basting stitch, then remove the painter's tape around the perimeter, leaving the tape on the Fabric 2 strips in place.

6. Due to the floating strips in this design, you will need to quilt the panel in order to keep it together in the long run. If you don't, the strips will twist and shift. There is no need to add batting or fleece, simply sew through the strips and interfacing. Quilt as desired. I quilted along the edge of the design created by the Fabric 1 strips, then removed the painter's tape. I then sewed along the edge of the design created by the Fabric 2 strips. Using matching thread helped hide any imperfections in the quilting. Using a milliners needle and perle cotton, I added a big stitch quilting "X" in the center of any gaps between strips. *fig C*

A

B

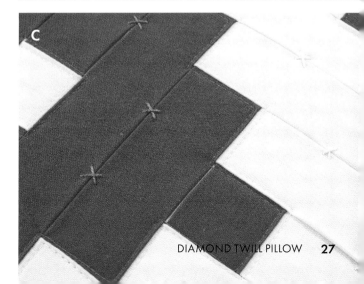

C

backing

1. Fuse the fusible fleece to the 23″ backing fabric square. Quilt as desired. I used painter's tape to sew vertical lines, then placed the tape horizontally and sewed along them to create a quilted grid. *fig D*

2. Cut off a 4″ × 23″ strip.

3. Center the zipper along the long side of the 4″ strip so that the zipper pull is facing the right side of the fabric, and the edge of the zipper tape lines up with the edge of the backing strip. Use clips or pins to hold the zipper in place. Use a zipper foot to sew on the zipper. You will need to move the zipper pull by unzipping the zipper in order to finish sewing. Fold back the backing strip and press, then top stitch along the fold.

4. Repeat this process to attach the other side of the zipper to the 19″ × 23″ piece of quilted backing. *fig E*

Make sure to move the zipper pull when sewing, and don't forget to top stitch. *fig F*

5. Trim the zipped backing to 19½″ × 19½″ so that it is the same size as the woven panel.

G

layering and binding

LAYERING

Lay the woven panel and backing wrong sides together. Pin or clip around all four sides. Baste stitch using a scant quarter inch seam. *fig G*

SEWING THE BINDING

Follow the general instructions at Binding, Steps 1 and 2 (page 10) to join your binding strips and press to get 1¼″ single-fold binding.

ATTACHING THE BINDING

Lay your pillow cover down, woven side up. Align the raw edge of one end of your binding strip against the middle of a raw edge of the pillow cover. Follow the instructions at Binding, Steps 3–10 (page 10) to attach the binding to the pillow.

finishing

Now it's time to stuff your pillow with the pillow form. I love a super over-stuffed cushion, so I made my own pillow form, measuring 21″ × 21″ and stuffed with recycled shredded foam. I sewed two 22″ × 22″ pieces of muslin right sides together using a generous ¼″ seam, leaving a hole for turning and stuffing. After stuffing, I hand sewed the opening closed.

This one-of-a-kind pillow will look great on your sofa or a favorite chair!

TARTAN WEAVE ON-POINT STORAGE BASKET

FINISHED SIZE: 8½″ × 8½″ × 9″

This woven tartan design is a weave that can work with both gradient and high-contrast color combinations. And pivoting the grid to a 45° diagonal literally turns a traditional idea into a modern project with dynamic color movement. Weave up one panel, then cut it in half to create this jaw-dropping storage basket.

TARTAN WEAVE BASKET
BY MATHEW BOUDREAUX

materials and supplies

MATERIALS

Fabric for weaving: 9 different 6″ × width of fabric strips. Fat quarters (precut 18″ × 20″–22″ rectangles of fabric) will work, but some strips will have to be sewn together to reach the longer diagonals.

Lightweight woven fusible interfacing: ⅔ yard

Medium woven fusible interfacing: 11″ × 19″

Fabric for lining and base: ⅝ yard

Foam stabilizer: 19″ × 27″ (I use By Annie's Soft and Stable)

SUPPLIES

Basic tools (page 7)

½″ Sasher or bias tape maker

½″ thick foam core board

½″ WEFTY Needle

cutting

WEAVE FABRICS

• Cut 6 strips 1″ × width of fabric from each of the 9 fabrics. Label your fabrics 1-9 to keep track of them while weaving.

LINING/BASE FABRIC

• Cut 1 strip 19″ × width of fabric. Subcut into 1 rectangle 19″ × 11″ and 1 rectangle 19″ × 29″.

preparing the strips

Using the instructions on Preparing the Strips (page 14), press the 9 sets of strips ready for weaving.

preparing the board

1. Using a ruler and dark colored marker, draw an 18″ × 18″ square on the foam core board. Draw two diagonals connecting opposite corners of the square, creating a big "X" on the board. Working from the drawn diagonal out, draw lines parallel to both diagonals, each measuring 2″ apart. *fig A*

2. Lay the lightweight woven fusible interfacing *adhesive side up* over the grid. You should feel the bumpy side on the top! This is important because you don't want the bumpy side sticking to the board later. You want it to stick to the underside of the *woven strips* later. Pin each corner, inserting the pins at an extreme angle.

weaving

The basket weave pattern you are creating is made up of 2 layers. Each layer of your woven panel will use roughly half of each color's strips.

FIRST LAYER

Starting with the innermost strips, begin pinning your strips to the board at a 45° angle, leaving no gap between the strips as you lay them. Follow the layout below while pinning strips beginning on the left or right of the center diagonal. You will use all 9 colors of strips to fill the grid. From the center line moving in either direction, the order of fabric placement is as follows: 1, 2, 3, 4, 4, 3, 2, 1, 5, 6, 7, 8, 9, 9, 8, 7, 6, 5, 1, 2, 3, 4, 4, 3, 2

Since the strip is much longer than the square, trim it to size after pinning in place and use the rest of the strip in another area, saving shorter strips for corners of the square. *fig B*

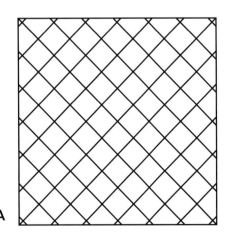

A

B

THREADING THE WEFTY

Thread your WEFTY by holding it with the WEFTY logo facing up and insert a strip into the eye of the needle, folding it back with raw edges together.

SECOND LAYER

The second layer of the woven panel is made up of the second set of strips in the same order as the first layer, woven in at a 90° angle.

1. Thread your WEFTY with the center strip and weave it into one corner of the first layer strips. Lift the first layer strips for the WEFTY using the Purple Thang.

THE WEAVING SEQUENCE YOU WILL FOLLOW IS UNDER ONE, OVER ONE, REPEATING EACH TIME. Ensure the strip is straight by peeking between the first layer strips to see the gridlines.

2. Pull taut, and then pin at the very end of the strip close to the outside edge of the first layer. Pin just outside of the first layer on the other corner and pin. Trim this strip just next to the pin.

3. Weave the next strip of Fabric 2 into the first layer right next to the strip you just wove in.

USE THE FOLLOWING SEQUENCE: OVER ONE, UNDER ONE, REPEATING EACH TIME.

4. Gripping each side of the strip, gently shimmy the strip back and forth to wiggle it closer to the first strip. Use a ruler and your Purple Thang to finish pushing the strip close to the first, making sure there are no gaps between them, then pin.

5. Keep alternating the weaving sequence and straightening the strips until you've finished weaving into the first layer. *fig C*

6. Carefully trim any loose threads.

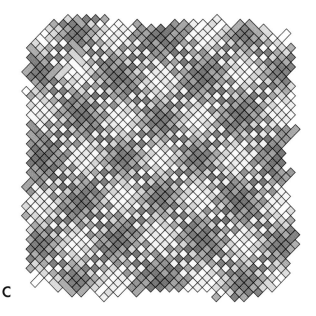

C

D

securing the weave

1. Press the panel gently with an iron on the steam setting, being mindful to avoid the pins. This will begin activating the fusible in the interfacing and securing woven strips.

2. Place painter's tape around the perimeter and finger-press firmly in place.

3. Remove the pins by holding the strip down with one finger while pulling out the pin. Make sure all pins are removed!

4. Press gently around the edges of the woven panel with the iron on the steam setting.

5. Carefully remove the panel from the board by gripping the sides of interfacing. Move the panel to your sewing machine and baste stitch using the edge of the tape as a guide.

6. Remove tape when finished. Trim ½″ outside of the basting stitch. *fig D*

E

sewing the basket

PREPARATION AND CUTTING

1. Measure 9½″ from one edge and mark a center line, dividing the woven square into two 9½″ × 19″ rectangles.

2. Tape on both sides of the marked line to secure each panel. *fig E*

3. Cut along the center line to create two panels, remove tape, and baste a scant ½″ away from the edge. *fig F*

F

OUTSIDE BASKET ASSEMBLY

1. Fuse interfacing to the 19″ × 11″ piece of lining/base fabric to create the bottom exterior panel.

2. Match the 19″ edge of interfaced panel with the long edge of one woven panel, right sides together, and sew using ½″ seam allowance (use throughout unless otherwise instructed). *fig G*

3. Press seam toward the base fabric and edge stitch along the fold of the base fabric. Trim off excess seam allowance. *fig H*

G

H

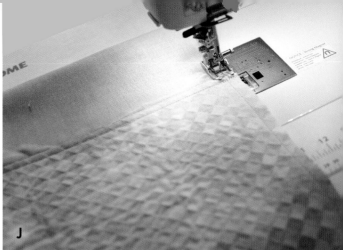

4. Repeat Steps 2 and 3 with the remaining woven panel. *fig I*

5. Fold this pieced 19″ × 27″ panel in half, with woven sections right sides together. At a minimum, clip or pin where the seams come together on both sides to make sure that they are aligned.

6. From the top, sew down both short edges of the folded piece. *fig J*

7. Using the stitch line and bottom fold as the edges, mark a 4¼″ square at both base fabric corners. Cut out the marked squares. *fig K*

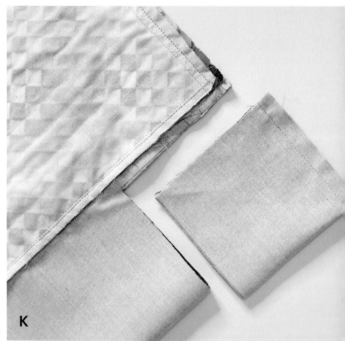

8. Rearrange one cut corner so that the side seam lies flat directly over the bottom fold, with raw edges aligned. The cut edge should form one long opening now. Stitch across the opening. Finish raw edges with a zigzag stitch, if desired. Repeat to box the remaining corner. *fig L*

9. Turn the basket right side out.

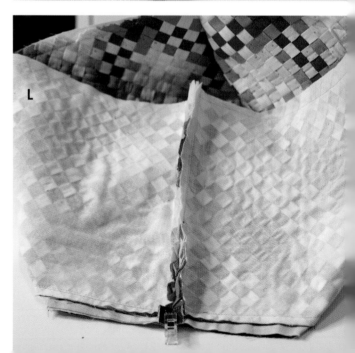

FOAM STABILIZER BASKET ASSEMBLY

Fold the 19″ × 27″ piece of foam stabilizer in half with 19″ edges aligned. Repeat Steps 6–8 of the outer basket assembly (page 36) to sew and box the corners, but DO NOT press the foam stabilizer. Trim excess seam allowances to just outside of the stitch line.

LINING ASSEMBLY

1. Fold the 19″ × 29″ piece of interfaced base fabric in half with 19″ sides matching.

2. From the top, unfolded edge, sew along both short ends of the piece with ½″ seam allowance.

3. Repeat Steps 7 and 8 of the outer basket assembly (page 36) to trim and box the corners. Do not turn lining right side out.

BASKET ASSEMBLY

1. Fold and press ¼″ of the top of the lining fabric toward the wrong side. Press another ½″ under to create faux binding. *fig M*

2. Stack the three basket layers: woven outer (right side out), foam stabilizer center, and lining (inside out), inside of each other. Take time to align seams carefully and clip them in place.

3. Position the faux binding fabric edge to cover the top, raw edge of the outside basket and the foam stabilizer. Clip in place and topstitch around the edge. *fig N*

Enjoy your basket and put pretty things in it.

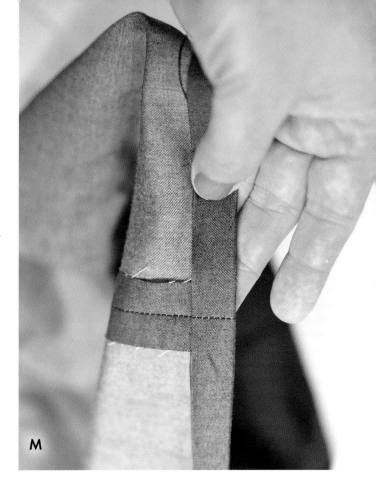

M

N

OMBRÉ MONOCHROME TOTE

FINISHED SIZE:
12″ × 13″ × 6″
(excluding handles)

The simplicity of a really well-done monochrome weave allows for the texture of the weave to take front and center. Add in using an ombré fabric and random patterns begin to emerge amongst the texture. This turns the simple into a truly one-of-a-kind tote great for any occasion or season.

OMBRÉ TOTE BY MATHEW BOUDREAUX

materials and supplies

MATERIALS

Fabric for weaving: 1¼ yard

Medium woven fusible interfacing (20″ wide): 2 yards

Fabric for lining and handles: 1⅛ yards

SUPPLIES

Basic tools (page 7)

1″ Sasher or bias tape maker

½″ thick foam core board

1″ WEFTY Needle

cutting

WEAVE FABRIC

• Cut 17 strips 2″ × width of fabric.

INTERFACING

• Cut 1 rectangle 18″ × 20″.

• Cut 1 rectangle 17″ × 26″.

• Cut 1 rectangle 8″ × 17″.

• Cut 2 rectangles 3″ × 17″.

LINING FABRIC

• Cut 2 strips 6″ × width of fabric (for handles). Trim to 40″ long.

• Cut 1 strip 17″ × width of fabric (for exterior bottom and bag lining). Subcut into 1 rectangle 8″ × 17″, and 1 rectangle 17″ × 26″.

• Cut 1 strip 3″ × width of fabric (for exterior top, and bag lining). Sub cut into 2 rectangles 3″ × 17″

preparing the strips

Using the instructions on Preparing the Strips (page 14), press 17 beautiful strips ready for weaving.

preparing the board

1. Using a ruler and dark colored marker, draw a 16″ × 18″ rectangle on the foam core board. Fill this in with a 2″ grid by drawing lines every 2″ horizontally and vertically. *fig A*

2. Lay the 18″ × 20″ medium woven fusible interfacing *adhesive side up* over the grid. You should feel the bumpy side on the top! This is important because you don't want the bumpy side sticking to the board later on. You want it to stick to the underside of the *woven strips* later. Pin each corner, inserting the pins at an extreme angle.

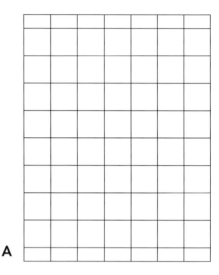

A

weaving

The basket weave pattern you are creating is made up of 2 layers.

FIRST LAYER

Lay a fabric strip down on the board, raw edges facing down. You should be able to see the gridlines through the interfacing. Use them as guides to keep the strip straight. Pull the strip taut and pin. Trim. Use the excess for the next strip, until your strip is no longer long enough. Repeat until your grid is covered. Make sure strips are right next to one another without overlapping. You don't want to be able to see the interfacing between the strips. *fig B*

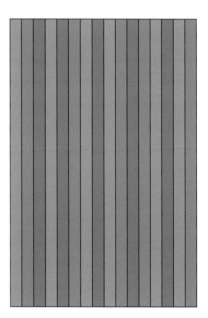

B

THREADING THE WEFTY

Thread your WEFTY by holding it with the WEFTY logo facing up and inserting a strip into the eye of the needle, folding it back with raw edges together.

SECOND LAYER

The second layer of the woven panel is made up of the same fabric, woven into the first layer at a 90° angle.

1. Thread your WEFTY with a strip and weave it into one side of the first layer strips. Lift the first layer strips for the WEFTY using the Purple Thang.

THE WEAVING SEQUENCE YOU WILL FOLLOW IS UNDER ONE, OVER ONE, REPEATING. Ensure the strip is straight by peeking between the first layer strips to see the gridlines.

2. Pull taut, and then pin at the very end of the strip close to the outside edge of the first layer. Pin just outside of the first layer on the other side of the last Layer 1 strip. Trim this strip just next to the pin.

3. Weave the remainder of the strip into the first layer right next to the strip you just wove in.

USE THE FOLLOWING SEQUENCE: OVER ONE, UNDER ONE, REPEATING.

4. Gripping each side of the strip, gently shimmy the strip back and forth to wiggle it closer to the first strip. Use a ruler and your Purple Thang to finish pushing the strip close to the first, making sure there are no gaps between them, then pin.

5. Keep alternating the weaving sequence and straightening the strips until you've finished weaving into the first layer. *fig C*

6. Carefully trim any loose threads.

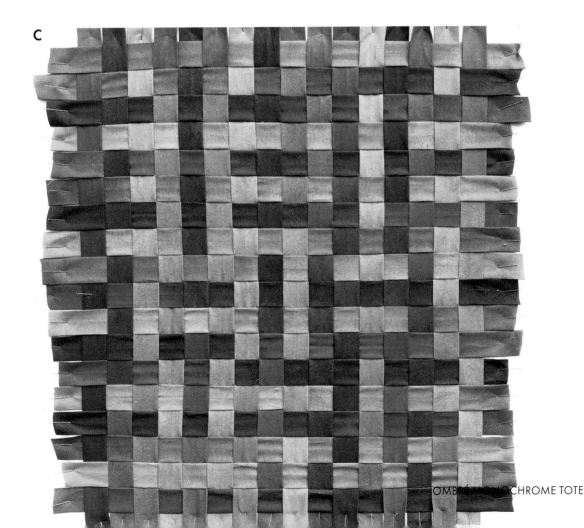

C

securing the weave

1. Press the panel gently with an iron on the steam setting, being mindful to avoid the pins. This will begin activating the fusible in the interfacing and securing the woven strips.

2. Place painter's tape around the perimeter and finger-press firmly in place.

3. Remove the pins by holding the strip down with one finger while pulling out the pin. Make sure all pins are removed!

4. Press gently around the edges of the woven panel with the iron on the steam setting.

5. Carefully remove the panel from the board by gripping the sides of the interfacing. Move the panel to your sewing machine and baste stitch using the edge of the tape as a guide. *fig D*

6. Remove tape when finished. Trim just outside of the basting stitch. Once trimmed, the woven panel will measure 16″ × 18″. *fig E*

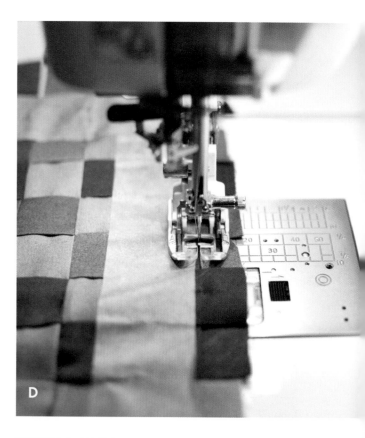

D

E

quilting the weave

Beginning with one of the upper corners, temporarily mark the 45° diagonal line across both edges of the panel. This line should cut each woven square into diagonal halves. From here, mark all diagonal seam lines parallel to the one you previously marked. Quilt over the marked lines using a walking foot and stitch length of 3 mm. Repeat the marking and quilting process once more for the second set of diagonal lines. *fig F*

the tote exterior

PREPARING THE STRAPS

1. Wrong sides together, fold one of the 6″ × 40″ strips lengthwise and press.

2. Now 3″ × 40″, open up the first fold and press the long raw edges in so that they meet in the middle of the wrong side.

3. Fold on the original pressed line, lining up the two folded edges, and press. The strap should now measure 1½″ wide.

4. Edge stitch at ⅛″ along both sides of the strap.

5. Follow the same process for the second strap.

ASSEMBLING THE EXTERIOR PANEL

1. Cut the woven panel in half and trim as needed to create two 8″ × 17″ rectangles. *fig G*

2. Fuse interfacing to the wrong sides of the 8″ × 17″ and two 3″ × 17″ lining fabric rectangles.

F

G

3. Right side up, position one of the 8″ × 17″ woven panels with the center cut toward the bottom. Place the end of one of the straps on the seam allowance of the fourth strip from the top left. Making sure not to twist the strap, place the other strap end on the seam allowance of the fourth strip from the top right. Baste strap ends in place.

4. Line up one of the 3″ × 17″ interfaced rectangles face down along the top raw edge of the woven panel and straps. Sew using a ½″ seam allowance along the top raw edge of all layers. *fig H*

5. Fold over the interfaced fabric and press, leaving strap facing down. Edge stitch the fabric fold. *fig I*

H

I

6. Fold each strap up and secure to the interfaced lining fabric with a ¾″ square with an "X" in the center. *fig J*

7. Repeat Steps 2–6 for the second woven side panel and strap.

8. In the same way, attach the interfaced 8″ × 17″ rectangle to the bottom of both side panels, keeping the straps out of the way of your stitching. *fig K*

J

K

CONSTRUCTING THE EXTERIOR

1. Fold the pieced 17″ × 26″ panel in half, with woven panels right sides together. At a minimum, clip/pin where the seams come together on both sides to make sure that they are aligned.

2. From the top, sew down both short edges of the folded piece, backstitching at the top. *fig L*

3. Using the stitch line and bottom fold as the edges, mark a 2½″ square at both fabric corners. Cut out marked squares. *fig M*

4. Rearrange one cut corner so that the side seam lies flat directly over the bottom fold, with raw edges aligned. The cut edge should form one long opening now. Stitch across the opening. Finish raw edges with a zigzag stitch, if desired. Repeat to box the remaining corner. *fig N*

5. Turn right side out.

L

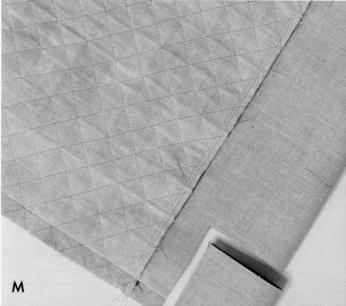

M

N

constructing the lining

1. Fuse 17″ × 26″ interfacing to the wrong side of the 17″ × 26″ lining fabric rectangle.

2. Repeat Steps 1–4 of Constructing the Exterior for the 17″ × 26″ interfaced lining fabric.

assembling the tote

1. Right sides together, insert the exterior into the lining and clip the raw edges together. Make sure to line up the two side seams. The straps will hang in between the two bags.

2. Leaving a 4″ gap, sew around the top edge using a ½″ seam allowance. *fig O*

3. Turn the tote right side out through the gap. Push the lining into the exterior and press.

4. Hand stitch or machine sew the 4″ turning gap.

Go find some stuff to put in your tote and enjoy it!

O

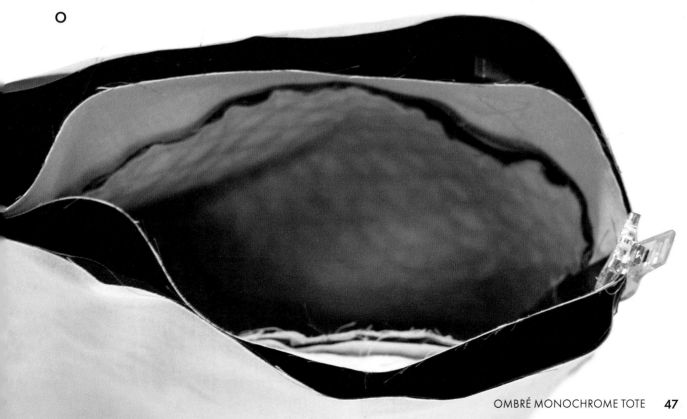

SMALL HOUNDSTOOTH CROSS-BODY BAG

FINISHED SIZE:
9″ × 10″

Select two fabrics with good contrast to highlight the fun houndstooth weave in this stylish cross-body bag.

CROSS-BODY BAG BY MATHEW BOUDREAUX

materials and supplies

MATERIALS

Fabric 1: ½ yard

Fabric 2: ½ yard

Lightweight woven interfacing: ¼ yard

SUPPLIES

Basic tools (page 7)

½˝ Sasher or bias tape maker

½˝ thick foam core board

½˝ WEFTY Needle

9˝ metal zipper

cutting

FABRIC 1

• Cut 1 bias binding strip 2˝ × 25˝ first.

• Cut 2 strips 4˝ × 30˝.

• Cut 1 rectangle 9˝ × 10˝.

• Cut 8 strips 1˝ × 11˝.

• Cut 10 strips 1˝ × 10˝.

FABRIC 2

• Cut 2 rectangles 9˝ × 10˝.

• Cut 8 strips 1˝ × 11˝.

• Cut 10 strips 1˝ × 10˝.

INTERFACING

• Cut 2 rectangles 9˝ × 10˝.

preparing the strips

Using the instructions on Preparing the Strips (page 14), press 18 beautiful strips ready for weaving from each color. Separate these out into the two lengths of Fabric 1 and Fabric 2 to keep them organized.

preparing the board

1. Using a ruler and dark colored marker, draw an 8″ × 9″ rectangle on the foam core board. Fill this in with a 1″ grid by drawing lines every 1″ horizontally and vertically. *fig A*

2. Lay 1 piece of lightweight woven fusible interfacing *adhesive side up* over the grid. You should feel the bumpy side on the top! This is important because you don't want the bumpy side sticking to the board later on. You want it to stick to the underside of the *woven strips* later. Pin each corner, inserting the pins at an extreme angle.

weaving

The basket weave pattern you are creating is made up of 2 layers. The first layer of your woven panel is made up of the 11″ strips of Fabrics 1 and 2.

FIRST LAYER

From left to right, the order of strips is 2, 1, 1, 2, 2, 1, 1, 2, 2, 1, 1, 2, 2, 1, 1, 2 *fig B*

Lay the first strip down on the board, raw edges facing down. You should be able to see the gridlines through the interfacing. Use them as guides to keep the strip straight. Pull the strip taut and pin. Trim. Repeat until your grid is covered. Make sure strips are right next to one another without overlapping. You don't want to be able to see the interfacing between the strips.

THREADING THE WEFTY

Thread your WEFTY by holding it with the WEFTY logo facing up and inserting a strip into the eye of the needle, folding it back with raw edges together.

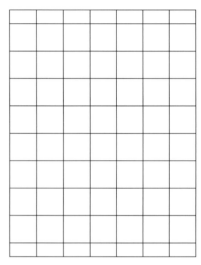

A

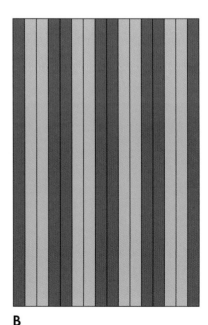

B

SECOND LAYER

From top to bottom, the order of strips is 2, 2, 1, 1, 2, 2, 1, 1, 2, 2, 1, 1, 2, 2, 1, 1, 2, 2, woven into the first layer at a 90° angle. Use the 10˝ strips of Fabrics 1 and 2.

1. Thread your WEFTY with a strip and weave it into one side of the first layer strips. Lift the first layer strips for the WEFTY using the Purple Thang.

THE WEAVING SEQUENCE YOU WILL FOLLOW IS UNDER ONE, OVER ONE, REPEATING. Ensure the strip is straight by peeking between the first layer strips to see the gridlines.

2. Pull taut, and then pin at the very end of the strip close to the outside edge of the first layer and trim. Pin just outside of the first layer on the other side of the prior strip. Trim this strip just next to the pin.

3. Weave the next strip into the first layer right next to the strip you just wove in.

USE THE FOLLOWING SEQUENCE: OVER ONE, UNDER ONE, REPEATING.

4. Gripping each side of the strip, gently shimmy the strip back and forth to wiggle it closer to the first strip. Use a ruler and your Purple Thang to finish pushing the strip close to the first, making sure there are no gaps between them, then pin. *fig C*

5. Keep alternating the weaving sequence and straightening the strips until you've finished weaving into the first layer. *fig D*

6. Carefully trim any loose threads.

C

D

E

securing the weave

1. Press the panel gently with an iron on the steam setting, being mindful to avoid the pins. This will begin activating the fusible in the interfacing and securing woven strips.

2. Remove the pins by holding the strip down with one finger while pulling out the pin. Make sure all pins are removed!

3. Press gently around the edges of the woven panel with the iron on the steam setting.

4. Carefully remove the panel from the board by gripping the sides of interfacing. Move the panel to your sewing machine and baste stitch next to the outer strips around all 4 edges.

5. Trim ⅜″ outside of the basting stitch and set aside with Fabric 1 and Fabric 2 9″ × 10″ rectangles. *fig E*

sewing the strap

1. Sew the two 4″ × 30″ strips right sides together at one of the short ends, using a ½″ seam allowance. Press the seam open.

2. Fold the now 4″ × 59″ strip lengthwise, wrong sides together, and press.

3. Open. Fold and press each 59″ raw edge to the centerline. *fig F*

F

4. Fold over along the original center crease, lining up the two folded edges, and press.

5. Edge stitch at ⅛″ along both sides of the strap. *fig G*

6. Trim two 3″ pieces from the strap and set aside for main zipper installation.

7. Press the 9″ × 10″ interfacing rectangle to the wrong side of the Fabric 1 rectangle. Baste an end of the strap to the right side of the interfaced rectangle 1″ from each top corner, making sure that the strap is not twisted. This is the back panel. *fig H*

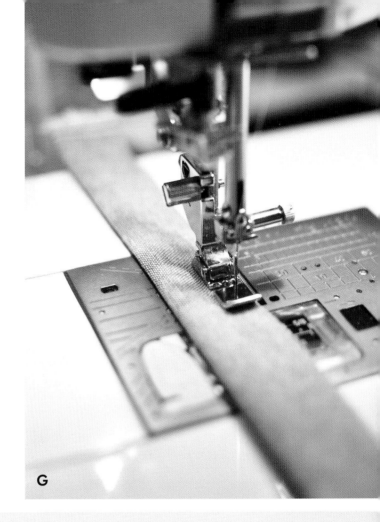

G

H

I

J

K

main zipper installation

1. Starting with one of the 9″ × 10″ Fabric 2 rectangles facing up, line up the edge of the zipper to the raw edge of a 9″ side of the rectangle. Make certain that the zipper pull is facing up. Place the back panel on top, right side down, also lining up to the same raw edge. Clip, pin, or glue these layers in place. *fig I*

2. Sew these layers together on the top using a ¼″ seam allowance. Use a zipper foot if needed. Flip the lining over the zipper so that both rectangles are wrong sides together and press the seam.

3. With the second Fabric 2 rectangle facing up, align with the remaining side of the zipper and the woven front panel and follow the same process you used for the back panel. *fig J*

binding the bag

1. Fold a 1″ × 3″ strap trimming lengthwise and pin over the raw zipper edge on each end. An inch of the strip should extend onto both the front and back sides of the bag. *fig K*

2. Edgestitch both strips in place. *fig L*

L

3. Close the panels onto each other right sides out. Baste in place around the raw edges. Use the curve template (page 142) to cut out a curve on both bottom corners. *fig M*

4. Fold the short ends of the 2″ × 25″ bias binding strip over ¼″to the wrong side and press. Fold the binding in half lengthwise, wrong sides together.

5. Line up and clip the raw edge of the binding with the raw edge of the woven panel. Both ends of the binding need to be basted in place where the end overlaps the zipper strips. The binding might feel too short to go around the perimeter of the bag, but since the binding was cut on the bias, it will stretch with you as you sew. You may end up with more than ¼″ when you get to the end of the binding strip! *fig N*

N

6. Sew the binding to the front of the bag using a ⅜″ seam allowance. *fig O*

7. Once the binding is sewn around the raw edges of the front, flip the binding toward the back of the bag. Press from the front to flatten the front binding, then press the fold and back before clipping in place. You can either choose to hand stitch the binding or machine edgestitch it.

O

LARGE HOUNDSTOOTH ZIPPED CASE

FINISHED SIZE: 13″ × 9″ × 2″

Once you realize that you can do even more woven designs when you go over or under multiple strips at a time, the sky's honestly the limit with potential. Maintaining an over-two-under-two repetition, the large houndstooth design is perfect for this classic zipped case.

ZIPPED CASE BY MATHEW BOUDREAUX

materials and supplies

MATERIALS

Fabrics for case: 1 yard each of 2 different colors (Fabrics 1 and 2)

Medium woven fusible interfacing: ⅔ yard

Clear vinyl such as Premium Clear Vinyl (C&T Publishing): ½ yard

Foam stabilizer: 13″ × 20″ (I used Bosal In R Form Plus double-sided fusible)

SUPPLIES

Basic tools (page 7)

½″ Sasher or bias tape maker

½″ thick foam core board

½″ WEFTY Needle

35″ zipper, for outer case closure

10″ zipper, for upper left pocket

cutting

FABRIC 1

• Cut 15 strips 1″ × width of fabric.

• Cut 1 rectangle 13½″ × 20½″ for lining.

• Cut at least 5 bias strips 2½″ wide from the remaining piece. Join to form a bias binding strip that is at least 70″ long.

FABRIC 2

• Cut 15 strips 1″ × width of fabric.

• Cut 1 rectangle 10″ × 17″. Subcut into 3 strips 2″ × 10″ (for upper left zipper pocket) and 1 rectangle 10″ × 11″ (for bottom left pocket).

• From the remaining piece, cut 1 strip 2″ × 13½″ (to bind right vinyl pocket) and 1 strip 2½″ × 14½″ (for center band).

INTERFACING

• Cut 1 rectangle 13″ × 21″

• Cut 1 strip 1½″ × 13″ (for center band).

• Cut 1 rectangle 5½″ × 10″ (for bottom left pocket).

VINYL

• Cut 1 rectangle 5″ × 10″ (for upper left zipper pocket).

• Cut 1 rectangle 9″ × 13½″ (for right vinyl pocket).

preparing the strips

Using the instructions on Preparing the Strips (page 14), press 15 beautiful strips ready for weaving from each color. Separate these out into Fabric 1 and Fabric 2 and keep them organized. Subcut 12 strips of each color into 1 piece 1˝ × 21˝ and 1 piece 1˝ × the remaining WOF.

preparing the board

1. Using a ruler and dark colored marker, draw a 12˝ × 20˝ rectangle on the foam core board. Fill this in with a 2˝ grid by drawing lines every 2˝ horizontally and vertically. *fig A*

2. Lay the 13˝ × 21˝ piece of medium weight woven fusible interfacing *adhesive side up* over the grid. You should feel the bumpy side on the top! This is important because you don't want the bumpy side sticking to the board later on. You want it to stick to the underside of the *woven strips* later. Pin each corner, inserting the pins at an extreme angle.

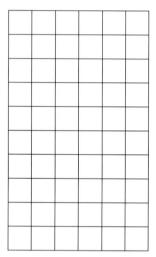

A

Weaving

The basket weave pattern you are creating is made up of 2 layers. The first layer of your woven panel is made up of 21˝ strips.

FIRST LAYER

From left to right, alternate between laying down groups of four strips of Fabric 1 and Fabric 2. Lay a Fabric 1 strip down on the board, raw edges facing down. You should be able to see the gridlines through the interfacing. Use them as guides to keep the strip straight. Pull the strip taut and pin. Trim. Repeat until your grid is covered. Make sure strips are right next to one another without overlapping. You don't want to be able to see the interfacing between the strips. *fig B*

THREADING THE WEFTY

Thread your WEFTY by holding it with the WEFTY logo facing up and inserting a strip into the eye of the needle, folding it back with raw edges together.

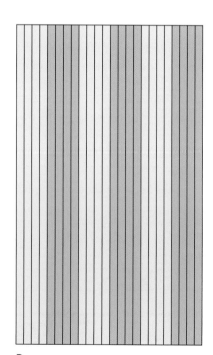

B

SECOND LAYER

The second layer of the woven panel is made up of 13″ strips which are woven into the first layer at 90° from right to left. Each strip will follow an under-two over-two pattern. There are four variations, however, of where the pattern begins. Repeat the following order from the bottom right side moving upward: *fig C*

Strip 1: under-two, over-two, under-two... until reaching the end.

Strip 2: under-one, over-two, under-two...

Strip 3: over-two, under-two, over-two...

Strip 4: over-one, under-two, over-two...

C

1. Thread your WEFTY with a Fabric 2 strip and weave it into one side of the first layer strips. Lift the first layer strips for the WEFTY using the Purple Thang.

THE WEAVING SEQUENCE YOU WILL FOLLOW IS UNDER TWO, OVER TWO, REPEATING. Fabric strips will be grouped into multiples of four, much like layer 1. Ensure the strip is straight by peeking between the first layer strips to see the gridlines.

2. Pull taut, and then pin at the very end of the strip close to the outside edge of the first layer. Pin just outside of the first layer on the other side of the Fabric 2 strip and pin. Trim this strip just next to the pin.

3. Weave the excess of the Fabric 2 strip (if there is 13″ left) or another Fabric 2 strip into the first layer right next to the strip you just wove in.

4. Gripping each side of the strip, gently shimmy the strip back and forth to wiggle it closer to the first strip. Use a ruler and your Purple Thang to finish pushing the strip close to the first, making sure there are no gaps between them, then pin. *fig D*

D

5. Keep following the weaving sequence, alternating between colors every 4 rows, and straightening the strips until you've finished weaving into the first layer. *fig E*

6. Carefully trim any loose threads.

securing the weave

1. Press the panel gently with an iron on the steam setting, being mindful to avoid the pins. This will begin activating the fusible in the interfacing and securing woven strips.

2. Place painter's tape around the perimeter and finger-press firmly in place. *fig F*

3. Remove the pins by holding the strip down with one finger while pulling out the pin. Make sure all pins are removed!

4. Press gently around the edges of the woven panel with the iron on the steam setting.

5. Carefully remove the panel from the board by gripping the sides of interfacing. Move the panel to your sewing machine and baste stitch using the edge of the tape as a guide.

6. Remove tape when finished. Trim ½˝ outside of the basting stitch.

E

F

sewing the interior

UPPER LEFT ZIPPER POCKET CONSTRUCTION

1. Gather the following components: 10″ zipper, 3 Fabric 2 strips each 2″ × 10″, and the 5″ × 10″ vinyl rectangle.

2. Right sides together, line up the edge of the 10″ zipper with the raw edge of one of the 2″ × 10″ rectangles. Secure with tape, if desired. Using a zipper foot, stitch ¼″ from the teeth of the zipper.

3. Carefully press fabric away from the zipper edge. Edge stitch ⅛″ from the fold. *fig G*

4. Stack one of the 2″ × 10″ Fabric 2 strips on top of the 5″ × 10″ vinyl, matching one long edge. Sew in place. Press the fabric away from the vinyl. Use low heat and avoid the vinyl as much as possible to keep it from melting.

5. Fold the fabric over (towards the back of vinyl) so that it extends beyond the seam. Press. Fold fabric in half lengthwise once more (encasing the seam), and press.

6. Place the zipper right side up. Align the fabric edge of the vinyl pocket made in Steps 4 and 5 along the bottom half of the zipper. Pin or use tape to secure, and then sew in place. *fig H*

7. Fold the 13½″ × 20½″ lining fabric rectangle in half both vertically and horizontally, finger-pressing both midlines. This will be the interior of the case. The fold lines will help in aligning the pockets.

8. Place the zipper pocket in the upper left quadrant, matching edges with the creases marking midlines. Using a long stitch length, baste in place across the top of the pocket. *fig I*

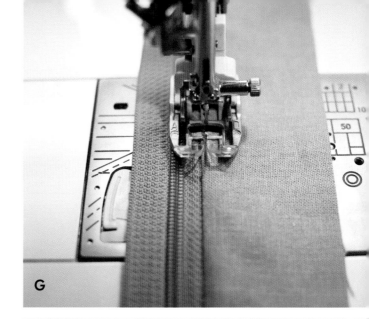

G

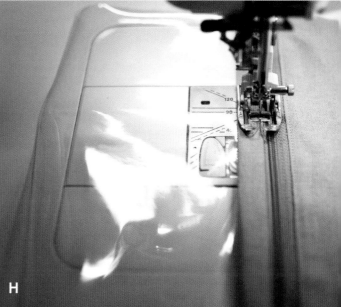

H

I

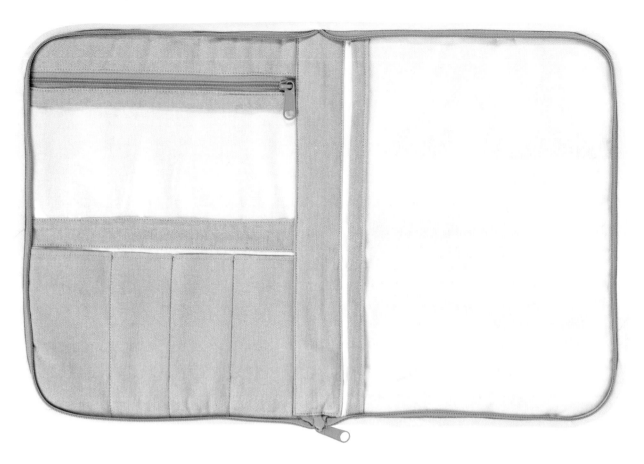

9. Press one long edge of the pocket band (the remaining 2″ × 10″ Fabric 2 rectangle) under by ½″. Align the unfolded edge with the bottom of the vinyl pocket, right sides together.

10. Sew the bottom edge of the pocket with ½″ seam allowance. Press the pocket band away from the vinyl.

11. Edge stitch the pocket band along both 10″ edges at ⅛″, attaching the pocket to the lining. *fig J*

12. Baste the remaining two sides of the pocket to the lining with a ¼″ seam allowance.

J

BOTTOM LEFT NOTIONS POCKET

1. Gather the following components: 10″ × 11″ Fabric 2 rectangle and 5½″ × 10″ interfacing rectangle.

2. Align the interfacing with one 10″ edge of the wrong side of fabric, and fuse in place.

3. Fold the piece in half lengthwise, wrong sides together, lining up the raw 10″ edges. Press and edge stitch along the fold. *fig K*

4. Place the pocket on the bottom left corner of the panel, aligning raw edges. Mark vertical lines at 3″, 5″, and 7″ from the left edge of the pocket. Stitch over each marked line, from bottom to top edge, pivoting at top edge and reinforcing the seam as you sew back down to the bottom of the pocket. *fig L*

5. With a long stitch length, baste the side and bottom edges of the pocket in place.

LARGE VINYL POCKET

1. Gather the 9″ × 13½″ vinyl rectangle and the 2″ × 13½″ Fabric 2 pocket binding. Right sides together, align the pieces along a 13½″ edge. Stitch together using a ½″ seam allowance.

2. Fold the fabric away from the vinyl and press seam toward the fabric.

3. Double fold fabric over the raw edge of the vinyl (encasing it) and press.

4. Topstitch ⅛″ from the fold. You'll attach this pocket to the lining after adding the foam stabilizer.

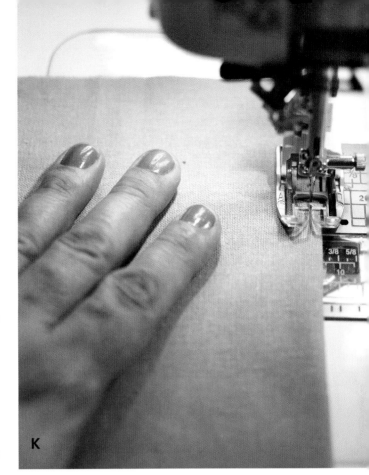

K

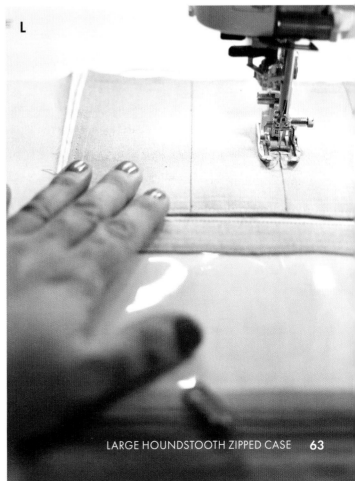

L

PREPARE THE CENTER BAND

1. Gather the 2½″ × 14½″ fabric strip and the 1½″ × 13″ interfacing. Match center of interfacing and fabric strip along one short edge. Fuse in place.

2. Fold both longer raw edges toward the wrong side ½″ from each edge. Press. Fold the non-interfaced short end of the center band toward the wrong side, 1″ from the raw edge. Press. *fig M*

ADD FOAM STABILIZER

1. Fuse foam stabilizer to the wrong side of the interior lining panel. If using a sew-in stabilizer, baste stabilizer in place all around the outer edge.

2. Align the raw edges of the large vinyl pocket with the right half of the interior panel. Baste around all 3 edges, ¼″ in from the edge.

3. With the folds facing down, place the center band on the vertical midline of the interior panel, placing the folded short end ½″ from the bottom lining edge. Use tape or pin in place, then topstitch the two long edges ⅛″ from each edge. Make sure to not stitch the bottom fold. *fig N*

M

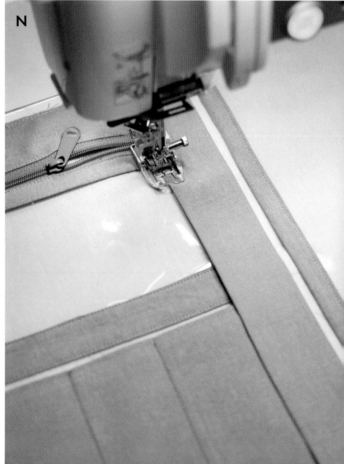

N

finishing the case

1. Stack the interior and exterior panel, wrong sides together. Fuse or baste in place around the edges. *fig O*

2. Using the curve template (page 142), draw each rounded corner. Baste just inside each drawn line and then cut along the curve. Trim sides as needed so all edges are flush. *fig P*

3. Unzip the 35″ zipper. With the case interior face up, insert the bottom end of the zipper face down in the bottom gap of the center band and top stitch in place across the bottom center band. *fig Q*

O

P

Q

Fold the right half of the zipper so that the tape edge of the zipper aligns with the case edge, and clip in place all around the case edge. Clip the zipper tape with scissors, but not more than ⅛″ deep, to help ease it around the corners. Repeat with the left half of the zipper and left case edge. *figs R–S*

4. Fold excess zipper length pointing away from the case.

5. Using a zipper foot and ¼″ seam allowance, sew the zipper in place, backstitching over zipper ends. Trim away excess zipper length. *fig T*

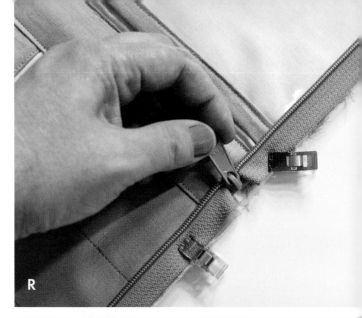

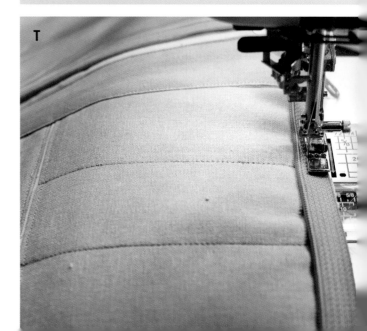

binding

1. Fold the 70˝ bias strip in half lengthwise, with wrong sides together. Press.

2. Follow the instructions at Binding, Steps 3–10 (page 10) to attach the binding to the case. Leaving the first few edges of binding hanging free, align the raw edges of the bias strip with the right edge of the interior of the case, pinning or clipping the binding in the middle of the lower right edge.

3. Sew the bias strip around the edge of the case with a ⅜˝ seam allowance. When sewing over the center bottom, make sure to avoid sewing over the zipper end. Stop sewing before you reach the starting point, leaving several inches of bias tape hanging free. *fig U*

4. Join the two short ends of the binding.

5. Fold the binding toward the outside of the case, clip, and hand sew in place.

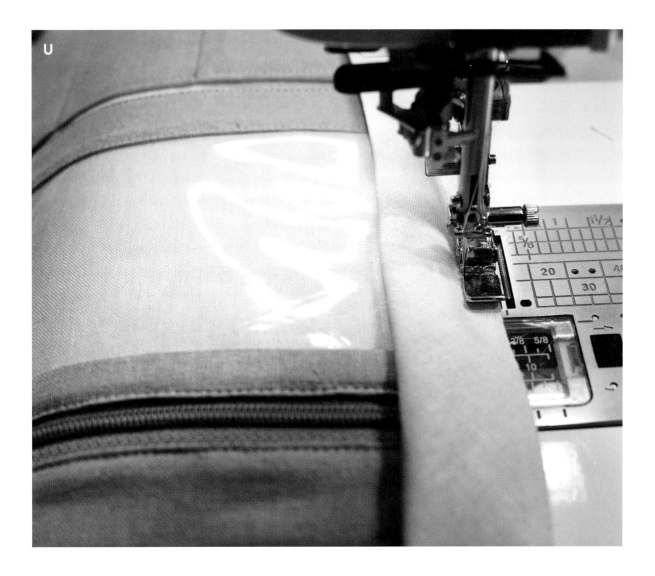

SILLY SILO HOOP ART

FINISHED SIZE: 14″ hoop

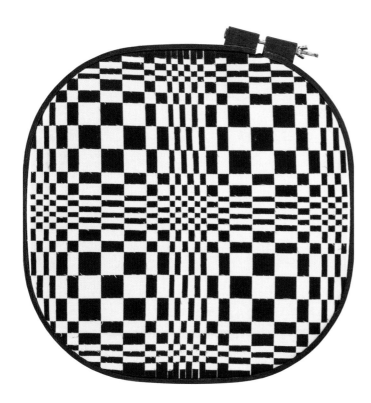

Using fabric reinforced with two-sided webbing, mix the width of strips and play with their placement to create some fun effects. Create an optical illusion and display it in a customized wooden embroidery hoop!

SILLY SILO WEAVE BY TARA CURTIS/WEFTY, MADE USING TWO CONTRASTING BELLA SOLIDS FROM MODA.

materials and supplies

MATERIALS

Fabric for weaving: ⅓ yard each of two contrasting fabrics (Fabric 1 and Fabric 2)

Muslin: 1 fat quarter (18″ × 22″)

Fusible two-sided webbing (I use Misty Fuse because I don't like to deal with removing paper)

SUPPLIES

Basic tools (page 7)

½″ thick foam core board

1″ and ½″ WEFTY Needles

14″ wooden embroidery hoop

40-60 grit sandpaper

Silpat mat or parchment paper

Acrylic craft paint (I used FolkArt Color Shift Acrylic Paint)

Acrylic paint brushes (I used Royal & Langnickel Menta Synthetic Acrylic 5 Piece Brush Set)

Adhesive (I used Aileen's tacky glue and Aileen's Fabric Fusion)

cutting

NOTE *Reinforce your weave fabrics with webbing using the manufacturer's instructions and either your silpat mat or parchment paper before cutting your weave strips. This will help mitigate some of the fraying.*

WEAVE FABRICS

Trim selvages and cut all strips along the grainline.

• Cut 14 strips ¼″ × 18″ from both Fabric 1 and Fabric 2.

• Cut 8 strips ½″ × 18″ from both Fabric 1 and Fabric 2.

• Cut 6 strips 1″ × 18″ from both Fabric 1 and Fabric 2.

preparing the board

1. Using a ruler and dark colored marker, draw a 16″ × 16″ square on the foam core board. Fill this in with a 2″ grid by drawing lines every 2″ horizontally and vertically. Or, you can use the 2″ grid you drew for the Diamond Twill Pillow (page 24) *fig A*

2. Lay the muslin over the grid. Pin each corner, inserting the pins at an extreme angle.

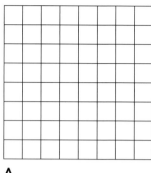

A

weaving

The basket weave pattern you are creating is made up of 2 layers. The first layer of your woven panel is made up of Fabric 1.

FIRST LAYER

Take care to lay strips down with the fusible side facing the muslin. Lay them in the following order:

> Four ¼″ strips, two ½″ strips, three 1″ strips, two ½″ strips, six ¼″ strips, two ½″ strips, three 1″ strips, two ½″ strips, and four ¼″ strips.

THREADING THE WEFTY

You can use the WEFTY to weave in the ¼″ wide strips by threading it further up the needle in the smaller slots. *fig B*

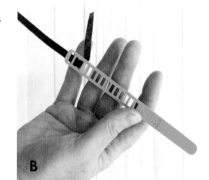

B

SECOND LAYER

1. The second layer of the woven panel is made up of Fabric 2, woven into the first layer at 90°. Follow the same pattern of strip sizes as in the first layer.

THE WEAVING SEQUENCE YOU WILL FOLLOW IS UNDER ONE, OVER ONE, REPEATING. Ensure the strip is straight by peeking between the first layer strips to see the gridlines.

2. Pull taut, and then pin in place. Pin at the very end of the strip close to the outside edge of the first layer and trim.

3. Weave the second strip. **THE WEAVING SEQUENCE YOU WILL FOLLOW IS OVER ONE, UNDER ONE, REPEATING.** Gripping each side of the strip, gently shimmy the strip back and forth to wiggle it closer to the first strip. Use a ruler and your Purple Thang to finish pushing the strip close to the first, making sure there are no gaps between them, then pin.

4. Keep alternating the weaving sequence and straightening the strips until you've finished weaving into the first layer. *fig C*

5. Carefully trim any loose threads from the fraying of the strips. A lint roller can also help with this step.

securing the weave

1. Press the panel gently with a hot dry iron, being mindful to avoid the pins. This will begin activating the fusible webbing on the strips to get them to stick to the muslin.

2. Remove the pins by holding the strip down with one finger while pulling out the pin. Make sure all pins are removed!

3. Press gently around the edges of the woven panel with the hot dry iron.

4. Carefully remove the panel from the board by gripping the sides of the muslin. Move the panel to your ironing board and press firmly to permanently adhere the woven panel to the muslin.

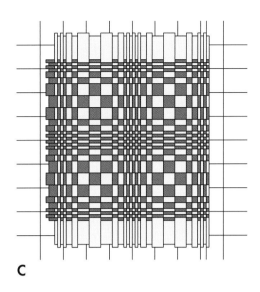

C

preparing the hoop

If you like, you can paint the wooden embroidery hoop to match your fabrics.

1. Sand the hoop with the sandpaper.

2. Prime it with white acrylic craft paint.

3. Mix your colored paints to get the color you want and evenly coat the outer ring of the hoop. Allow to thoroughly dry between coats.

installation and mounting

1. Lay your woven panel down on a smooth flat surface, woven side up.

2. Slip the inside ring of the embroidery hoop underneath the woven panel and center it.

D

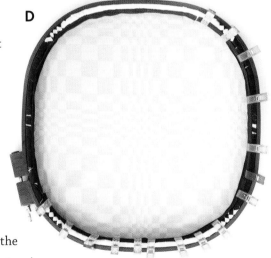

3. Making sure the outer ring is loosened completely, center it over the woven panel and inner ring and press down. Begin tightening the outer ring. You may need to adjust the woven panel as you tighten to avoid bunching.

4. Trim the excess fabric at the back of hoop, leaving two inches of excess. Apply a thin bead of glue to the inner ring, then wrap the excess around and press to adhere. *fig D*

5. You can use the ring of the hoop itself to hang the hoop, or else attach a short ribbon to the hardware on the outer ring and hang it by the ribbon.

Don't stare at the Silly Silo Weave for too long!

TRIAXIAL WEAVING

TUMBLING BLOCKS WOVEN COASTERS

FINISHED SIZE:
5½″ × 5½″ (6 coasters)

Tumbling blocks is a perfect weave to learn first when wanting to take on triaxial design. Weave up one panel, then cut it into squares to create these gorgeous and unique coasters.

TRIAXIAL WOVEN COASTERS BY TARA CURTIS/ WEFTY, MADE USING THREE GRADIENTS OF GRUNGE FOR MODA FABRICS

materials and supplies

MATERIALS

Fabric for weaving: ⅜ yard each of 3 gradients

Lightweight woven interfacing: ⅓ yard

Backing fabric: 1 fat quarter (18″ × 22″) or ⅓ yard

Binding: ½ yard

SUPPLIES

Basic tools (page 7)

½″ Sasher or bias tape maker

½″ thick foam core board

½″ WEFTY Needle

cutting

WEAVE FABRICS

• Cut 12 strips 1″ × width of fabric from each fabric (medium gradient is Fabric 1, dark gradient is Fabric 2, and light gradient is Fabric 3).

BACKING FABRIC

• Cut 6 squares 7″ × 7″.

BINDING

• Cut 5 strips 2½″ × width of fabric.

preparing the strips

Using the instructions on Preparing the Strips (page 14), press 12 beautiful strips ready for weaving from each color. Fold each Fabric 1 (medium) strip in half and cut to get a total of 24 strips, measuring approximately 20″ long. Keep your strips organized.

preparing the board

1. Draw a 13″ × 18″ rectangle on your foam core board. Using the 30° mark on your ruler lined up against a 13″ side of the rectangle, draw 2 angled lines in a bright color as shown in red and green in the illustration at right. *fig A*

2. Use a dark color for the rest of your markings. Carefully position the ruler so that it rests vertically across the point where the two 30° lines intersect and draw a line down the center of the board.

3. Continue adding lines along the vertical and diagonal in 1″ intervals to create an isometric grid.

4. Lay the lightweight woven fusible interfacing *adhesive side up* over the grid. You should feel the bumpy side on the top! This is important because you don't want the bumpy side sticking to the board later. You want it to stick to the underside of the *woven strips* later. Pin each corner, inserting the pins at an extreme angle.

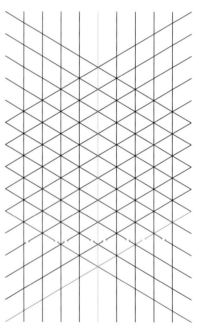

A

weaving

The triaxial weave pattern you are creating is made up of 3 layers. The first layer of your woven panel is made up of Fabric 1 (medium) strips.

FIRST LAYER

Lay a Fabric 1 strip down on the board, raw edges facing down. You should be able to see the gridlines through the interfacing. Use them as guides to keep the strip straight. Pull the strip taut and pin, ensuring you have at least 18˝ of space between the pins. Repeat until you have pinned 22 strips. You will have 2 strips leftover. *fig B*

THREADING THE WEFTY

Thread your WEFTY by holding it with the WEFTY lettering up and inserting a strip into the eye of the needle, folding it back with raw edges together. *fig C*

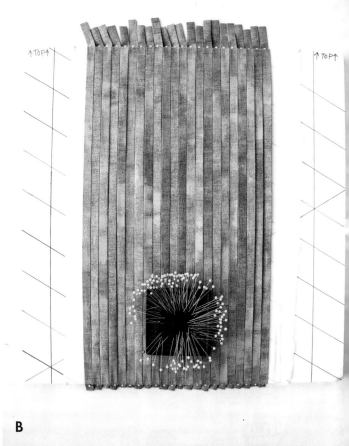

B

C

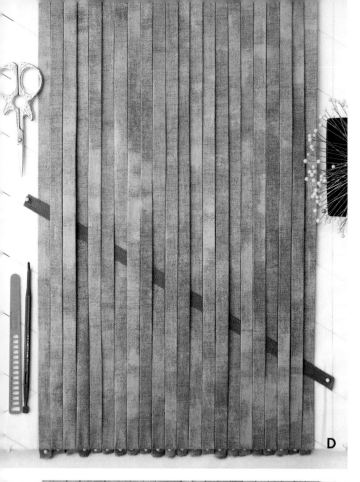

D

E

SECOND LAYER

The second layer of the woven panel is made up of Fabric 2 (dark), woven into the first layer, from right to left, at a 30° angle. Use That Purple Thang to hold up Fabric 1 strips for the WEFTY as you weave.

1. Working from the right, weave your first Fabric 2 strip into the first layer using the following sequence: **OVER ONE, UNDER TWO, REPEATING.**

Ensure that the strip follows along your brightly colored 30° line by peeking between the first layer strips to see the gridlines.

2. Pull taut, and then pin in place. Pin at the very end of the strip close to the outside right edge of the first layer. Then pin just outside of the first layer on the other left side. Trim the excess of this strip just next to the pin. *fig D*

3. Weave the excess of the Fabric 2 strip into the first layer right above the strip you just wove in.

WEAVE UNDER THE FIRST VERTICAL FABRIC 1 STRIP, THEN USE THE SEQUENCE OVER ONE, UNDER TWO, REPEATING.

4. Gripping each side of the strip, gently shimmy the strip back and forth to wiggle it closer to the first strip. Starting in the center of the strip, use That Purple Thang to finish pushing the strip close to the first, making sure there are no gaps between them, then pin. *fig E*

5. The third Fabric 2 strip is woven in under the first two vertical strips, then over one, under two, repeating. *fig F*

> **Tip: Disco Diamonds**
> *As you weave these strips in using the correct sequence, you will see that you are creating diamond shapes out of the Fabric 2 strips. The diamonds will touch tip-to-toe, all in a 30° row. Trisha Franklin of Quilt Chicken calls this phenomenon "Disco Diamonds" because each diamond's "toe" touches the "fingertip" of the diamond before it. Listen to some Saturday Night Fever and make sure your second layer of Fabric 2 strips is creating rows of diamonds that are tip-to-toe throughout!*

6. Repeat Steps 1–5 until the second layer covers the entire first layer, making sure to fill in the corners. *fig G*

When you reach the corners, you may need to temporarily unpin first layer strips before you can weave the rest of the way.

THREADING THE WEFTY

The WEFTY is designed with a tapered end to glide through several layers of folded fabric strips. To do this effectively, thread your WEFTY for the third layer differently than you have been by holding it with the numbers facing up.

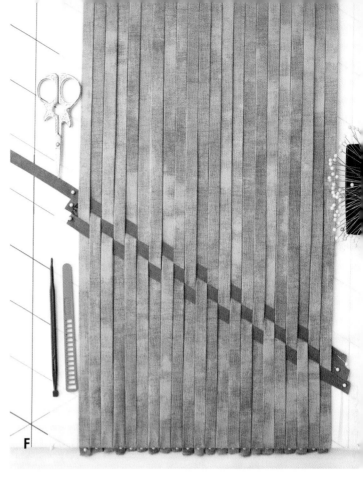

F

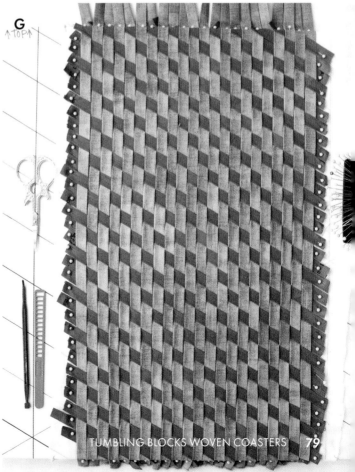

G
↑TOP↑

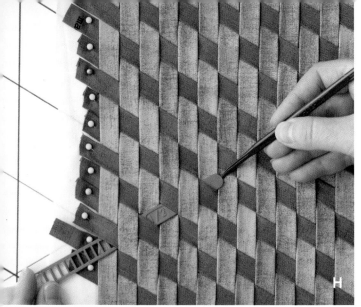

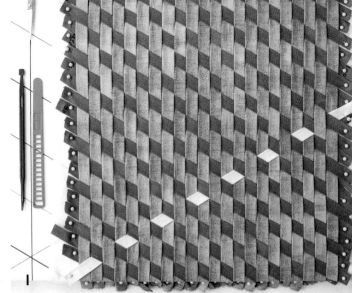

THIRD LAYER

The third layer is made up of Fabric 3 (light) strips and is woven in from the left side at a 30° angle. **THE SEQUENCE OF HOW IT IS WOVEN INTO THE FIRST LAYER IS UNDER ONE, OVER TWO.**

You may find the shape that the cluster of two second layer strips and one first layer strip makes: to some people it looks like a backwards letter "z" or a lightning bolt. When you begin weaving the third strip, start by weaving under one of those shapes. *fig H*

Use That Purple Thang by inserting it under the second layer strip on the right side of the shape and letting it act as a ramp for the WEFTY which is entering the left side of the shape.

In the photo above, Tara has started at the bottom left-hand corner to weave over two first layer strips, up under a second layer strip, under a first layer strip and out from under the next second layer strip. This action is repeated across the board. *figs I–K*

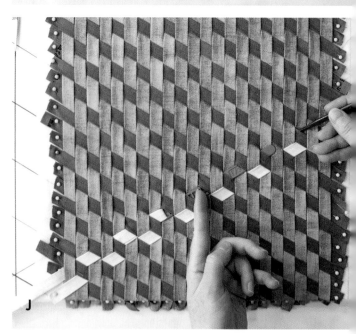

Your WEFTY and That Purple Thang should meet underneath the center of the first layer strip. You can't see them meet up under there, but you will hear a "click" if they find one another. If you feel resistance or notice the WEFTY getting caught in the strip folds, do not push through. Instead, back up, adjust, and try again.

Follow the arrows in the illustration at right to find the third layer placement. *fig L*

When you reach the corners, you will need to temporarily unpin the first two layers of strips to get the third layer strip woven in. *fig M*

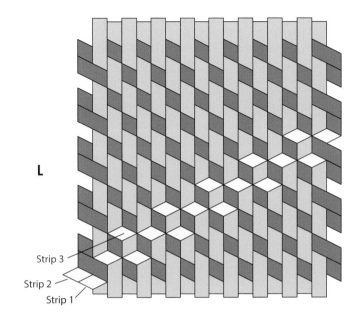

L

Strip 3
Strip 2
Strip 1

M

securing the weave

1. Press the panel gently with an iron on the steam setting, being mindful to avoid the pins. This will begin activating the fusible in the interfacing and securing woven strips.

2. Place painter's tape around the perimeter and finger-press firmly in place.

3. Remove the pins by holding the strip down with one finger while pulling out the pin. Make sure all pins are removed!

4. Carefully remove the panel from the board by gripping the sides of the interfacing. Move the panel to your sewing machine and baste stitch around all 4 edges using a ⅛″ seam.

NOW YOU WILL DIVIDE THE PANEL INTO SIX 5½″ SQUARES.

5. First find the center of the width of the panel. Measure from the baste stitches as opposed to measuring from the edge of your woven panel. Tape twice vertically, leaving a scant ¼″ space between.

6. Sew along the edge of each piece of tape. You will have two lines of stitching with a small space between. Peel off the strips of tape and discard. *fig N*

7. Now divide the woven panel crosswise into three 5½″ sections using two strips of tape and sewn lines as in Steps 5 and 6. You will divide each rectangle at 5½″, 11″, and 16½″. You will have a bit of excess from each rectangle. Discard or save for another project idea. *fig O*

8. Trim each square to 5½″ × 5½″ and baste stitch around the edges again if necessary.

N

O

layering and binding

LAYERING

Layer woven and backing squares wrong sides together. The backing fabric will extend beyond each woven side.

BINDING

Follow the general instructions at Binding, Steps 1 and 2 (page 10) to join your binding strips and press to get 1¼˝ single-fold binding.

ATTACHING THE BINDING

1. Lay your coaster sandwich down, woven side up. Align the raw edge of one end of your binding strip against the middle of a raw edge of the woven side. Follow the instructions at Binding, Steps 3–10 (page 10) to attach the binding to the pillow.

2. Repeat for the rest of the coasters. *fig P*

You're done! These look so nice around the house and make gorgeous gifts for teachers, co-workers, and sewing friends!

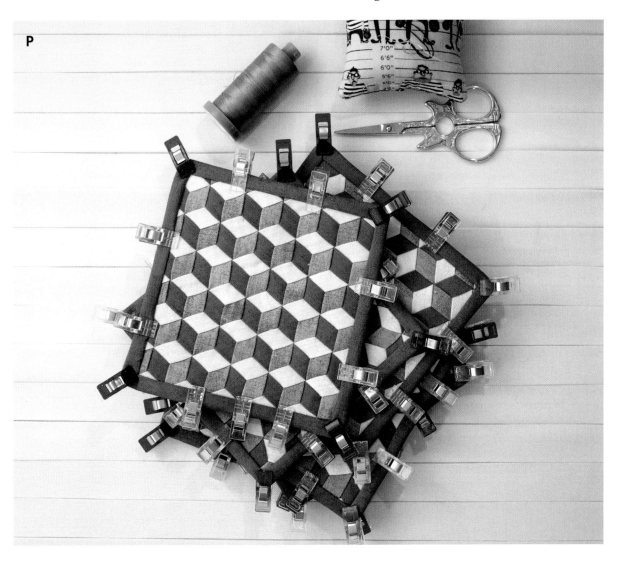

P

EVIL EYE WEAVE EVERYDAY ZIPPED POUCH

FINISHED SIZE: 6″ × 10″

Part of the fun of triaxial weaving is witnessing how minor adjustments to the order of placement can create next-level results. This Evil Eye Weave provides a classically innovative statement perfect for an Everyday Zipped Pouch.

EVIL EYE POUCH BY
MATHEW BOUDREAUX

materials and supplies

MATERIALS

Fabric 1 (turquoise): ½ yard

Fabric 2 (navy): ½ yard

Fabric 3 (gold): 1 fat quarter (18″ × 22″)

Lightweight woven interfacing: 12″ × 16″

SUPPLIES

Basic tools (page 7)

½″ Sasher or bias tape maker

½″ thick foam core board

½″ WEFTY Needle

11″ zipper

cutting

FABRIC 1

• Cut 1 rectangle 11″ × 13″ for lining first.

• From the remaining fabric, cut strips 2″ width of fabric. You should get 5 strips × about 27″ and 3 strips × about 40″

FABRIC 2

• Cut 6 strips 2″ × width of fabric.

FABRIC 3

• Cut 5 strips 2″ × 22″.

preparing the strips

Using the instructions on Preparing the Strips (page 14), press all strips ready for weaving from each color. Keep your strips organized.

preparing the board

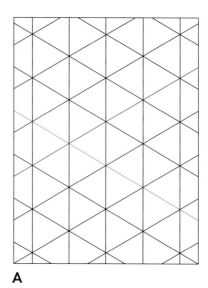

1. Draw a 10″ × 13″ rectangle on the foam board. Using the 30° mark on your ruler lined up along a 10″ edge of the rectangle, draw 2 angled lines in a bright color. *fig A*

2. Use a dark color for the rest of your markings. Carefully position the ruler so that it rests vertically across the point where the two 30° lines intersect and draw a line down the center of the board.

A

3. Continue adding lines along the vertical and diagonal in 2″ intervals to create an isometric grid.

4. Lay the lightweight woven fusible interfacing *adhesive side up* over the grid. You should feel the bumpy side on the top! This is important because you don't want the bumpy side sticking to the board later. You want it to stick to the underside of the *woven strips* later. Pin each corner, inserting the pins at an extreme angle.

weaving

The triaxial weave pattern you are creating is made up of 3 layers. The first layer of your woven panel is made up of Fabrics 1 and 2.

FIRST LAYER

From left to right, the order of the fabric strips is: 2, 1, 1, 2, 1, 1, 2, 1, 1, 2. Lay the first strip down on the board, raw edges facing down. You should be able to see the gridlines through the interfacing. Use them as guides to keep the strip straight. Pull the strip taut and pin, ensuring you have at least 18″ of space between the pins. Repeat until you have pinned all 10 strips. *fig B*

THREADING THE WEFTY

Thread your WEFTY by holding it with the WEFTY lettering up and inserting a strip into the eye of the needle, folding it back with raw edges together.

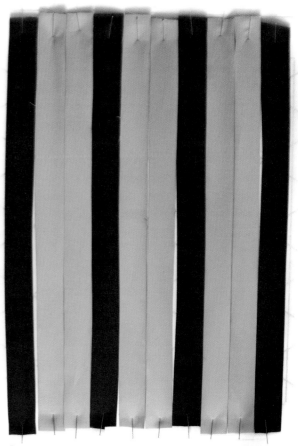

B

SECOND LAYER

The second layer of the woven panel is made up of Fabrics 1 and 2, woven into the first layer, from right to left, at 30°. Use That Purple Thang to hold up Fabric 1 strips for the WEFTY as you weave. The fabric order starting in the bottom right corner and moving upward is 2, 1, 1, 2, 1, 1, etc. The same order works moving downward until the grid is filled.

1. WORKING FROM THE RIGHT, WEAVE YOUR FIRST FABRIC 2 STRIP INTO THE FIRST LAYER USING THE FOLLOWING SEQUENCE: OVER ONE, UNDER TWO, REPEATING.

Ensure the strip follows along your brightly colored 30° line by peeking between the first layer strips to see the gridlines.

2. Pull taut, and then pin at the very end of the strip close to the outside right edge of the first layer. Then pin just outside of the first layer on the other left side. Trim the excess of this strip just next to the pin.

3. Weave a Fabric 1 strip into the first layer right above the strip you just wove in.

WEAVE UNDER THE FIRST VERTICAL FABRIC 2 STRIP, THEN USE THE SEQUENCE OVER ONE, UNDER TWO, REPEATING.

4. Gripping each side of the strip, gently shimmy the strip back and forth to wiggle it closer to the first strip. Starting in the center of the strip, use That Purple Thang to finish pushing the strip close to the first, making sure there are no gaps between them, then pin and trim the strip.

5. THE SECOND FABRIC 1 STRIP IS WOVEN IN UNDER THE FIRST TWO VERTICAL STRIPS, THEN OVER ONE, UNDER TWO, REPEATING. *fig C*

6. Repeat Steps 1–5 until the second layer covers the entire first layer, making sure to fill in the corners.

When you reach the corners, you may need to temporarily unpin some first layer strips before you can weave the rest of the way.

C

THREADING THE WEFTY

The WEFTY is designed with a tapered end to glide through several layers of folded fabric strips. To do this effectively, thread your WEFTY for the third layer differently than you have been by holding it with the numbers facing up

THIRD LAYER

The third layer is made up of Fabrics 2 and 3 strips and is woven in from the left side at a 30° angle. The fabric order starting in the bottom left corner and moving upward is 3, 2, 2, 3, 2, 2, etc. The same order works moving downward until you fill in the grid.

1. THE SEQUENCE OF HOW THE FIRST STRIP IS WOVEN INTO THE FIRST LAYER IS UNDER ONE, OVER TWO. Pull the strip taut, pin, and trim the excess.

2. Weave the second strip into the first layer right above the strip you wove in Step 1.

WEAVE OVER THE FIRST VERTICAL FABRIC 1 STRIP, THEN USE THE SEQUENCE UNDER ONE, OVER TWO, REPEATING.

3. THE THIRD LAYER 3 STRIP IS WOVEN IN OVER THE FIRST TWO VERTICAL STRIPS, THEN UNDER ONE, OVER TWO, REPEATING.

Use That Purple Thang by inserting it under the second layer strip on the right side of the shape and letting it act as a ramp for the WEFTY which is entering the left side of the shape.

D

Your WEFTY and That Purple Thang should meet underneath the center of the first layer strip. You can't see them meet up under there, but you will hear a "click" if they find one another. If you feel resistance or notice the WEFTY get caught in the strip folds, do not push through. Instead, back up, adjust, and try again.

When you reach the corners, you will need to temporarily unpin the first two layers of strips to get the third layer strip woven in. *fig D*

securing the weave

1. Press the panel gently with an iron on the steam setting, being mindful to avoid the pins. This will begin activating the fusible in the interfacing and securing woven strips.

2. Place painter's tape around the perimeter and finger-press firmly in place.

3. Remove the pins by holding the strip down with one finger while pulling out the pin. Make sure all pins are removed!

4. Press gently around the edges of the woven panel with the iron on the steam setting.

5. Carefully remove the panel from the board by gripping the sides of interfacing. Move the panel to your sewing machine and baste stitch using the outside edge of the tape as a guide.

6. Remove tape when finished. Trim ½" outside of the basting stitch. Quilt, if desired, before proceeding to the pouch construction. *fig E*

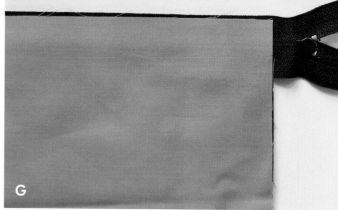

sewing the pouch

1. Stack the pouch pieces and zipper as follows: *figs F–G*

Woven panel, right side up

Zipper, right side down, centered along one 11″ edge of woven panel

Lining piece, right side down, aligned with edges of woven panel

Be sure that the zipper pull is entirely out of the path of stitching. Clip or pin pieces in place. Backstitch, then sew the zipper to the panel using a ⅜″ seam allowance. Backstitch. Fold the panel and lining over the seam allowance, press.

NOTE *Since the zipper is longer than required, you will not have to pause and adjust the zipper pull as you sew.*

2. Repeat Step 1 with the opposite zipper edge, the other 11″ panel end, and the opposite end of the lining piece. *figs H–I*

3. Using a ½″ seam allowance, and backstitching at the beginning and end of the seam, stitch the first side edge of the pouch with the zipper seams held toward the lining. *fig J*

Unzip the zipper halfway. Stitch the second side seam of the pouch, leaving a 2″ gap at the bottom corner of the lining. *fig K*

4. Turn the pouch right side out through the 2″ gap in the lining. Press and stitch the opening closed. Push the lining into the pouch, zip the pouch up, and enjoy having the most awesome pouch ever created in the history of pouches.

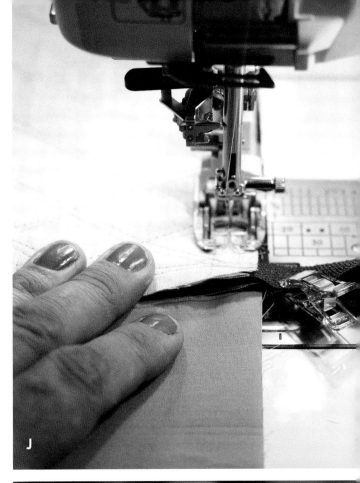

J

K

KALEIDOSCOPE WEAVE WALL ART

FINISHED SIZE: 9″ × 12″

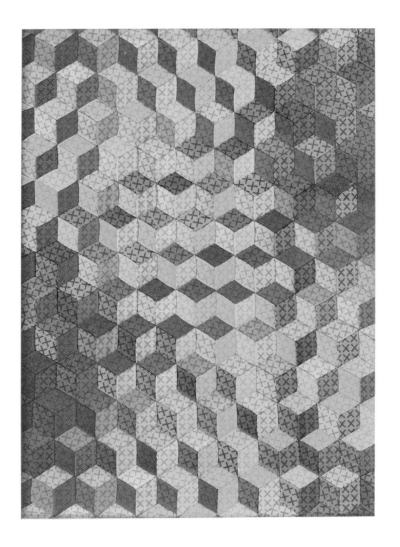

The kaleidoscope weave is definitely a next-level triaxial weave, especially this one, which has 18 different colors of fabrics. However, it's also super forgiving, so if you go out of order or skip one, it's still going to be gorgeous and no one will really know. If you're ready for this journey into the kaleidoscope, get ready to create a 9″ × 12″ framed piece of wall art that is sure to dazzle anyone who gazes upon its beauty.

KALEIDOSCOPE WALL ART BY
MATHEW BOUDREAUX

materials and supplies

MATERIALS

Fabric for weaving: 1 fat eighth (9″ × 22″) each of 18 different fabrics (this project is a great place to integrate scraps)

Lightweight woven interfacing: ⅓ yard

SUPPLIES

Basic tools (page 7)

½″ Sasher or bias tape maker

½″ thick foam core board

½″ WEFTY Needle

9″ × 12″ frame

cutting

WEAVE FABRICS

• Cut 5 strips 1″ × 22″ from each fabric (Fabrics 1 through 18).

preparing the board

1. Draw a 9″ × 12″ rectangle on the foam board. Using the 30° mark on your ruler lined up along a 9″ edge of the rectangle, draw 2 angled lines in a bright color as shown in red and green in the illustration at right. *fig A*

2. Use a dark color for the rest of your markings. Carefully position the ruler so that it rests vertically across the point where the two 30° lines intersect and draw a line down the center of the board.

3. Continue adding lines along the vertical and diagonal in 1″ intervals to create an isometric grid.

4. Lay the lightweight woven fusible interfacing *adhesive side up* over the grid. You should feel the bumpy side on the top! This is important because you don't want the bumpy side sticking to the board later. You want it to stick to the underside of the *woven strips* later. Pin each corner, inserting the pins at an extreme angle.

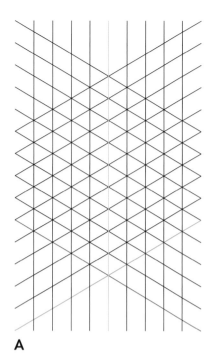

A

weaving

The triaxial weave pattern you are creating is made up of 3 layers. The first layer of your woven panel will use 1 strip of each color fabric for a total of 18 strips.

FIRST LAYER

Starting on the left of the grid, lay a Fabric 1 strip down on the board, raw edges facing down. You should be able to see the gridlines through the interfacing. Use them as guides to keep the strip straight. Pull the strip taut and pin, ensuring you have at least 13 inches of space between the pins. Repeat through all 18 fabrics until you have pinned 18 strips. Trim any excess and set aside. *fig B*

THREADING THE WEFTY

Thread your WEFTY by holding it with the WEFTY lettering up and inserting a strip into the eye of the needle, folding it back with raw edges together.

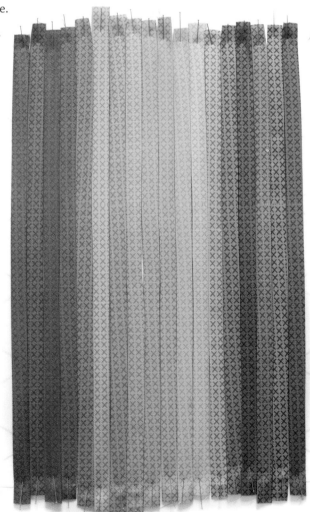

B

SECOND LAYER

The second layer of the woven panel uses the backward order of the fabrics from layer one, woven into the first layer, from right to left, at 30°. Use That Purple Thang to hold up Fabric 1 strips for the WEFTY as you weave.

The strip that begins on top of the bottom right corner is Fabric 15. As you move up, the fabric number will get smaller (14, 13, 12, 11, 10, etc.). As you move down, the Fabric number will increase until reaching 18, after which you start over again at 1 (16, 17, 18, 1, 2, etc.). Use That Purple Thang to hold up Fabric 1 strips for the WEFTY as you weave. *fig C*

1. WORKING FROM THE RIGHT, WEAVE YOUR FIRST LAYER 2 STRIP INTO THE FIRST LAYER USING THE FOLLOWING SEQUENCE: OVER ONE, UNDER TWO, REPEATING.

Ensure the strip follows along your brightly colored 30° line by peeking between the first layer strips to see the gridlines.

2. Pull taut, and then pin at the very end of the strip close to the outside right edge of the first layer. Then pin just outside of the first layer on the other left side. Trim the excess of this strip just next to the pin.

3. Weave the next layer 2 strip into the first layer right above the strip you just wove in.

WEAVE UNDER THE FIRST VERTICAL LAYER 1 STRIP, THEN USE THE SEQUENCE OVER ONE, UNDER TWO, REPEATING.

4. Gripping each side of the strip, gently shimmy the strip back and forth to wiggle it closer to the first strip. Starting in the center of the strip, use That Purple Thang to finish pushing the strip close to the first, making sure there are no gaps between them, then pin. *fig D*

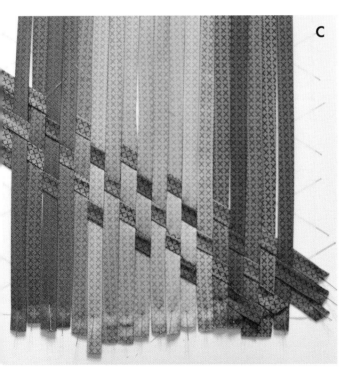

C

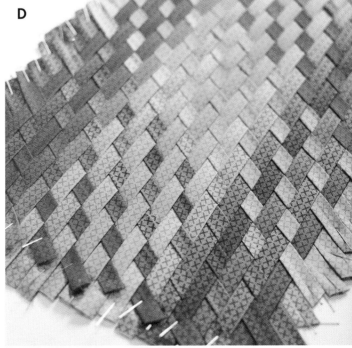

D

E

F

5. THE THIRD LAYER 2 STRIP IS WOVEN IN UNDER THE FIRST TWO VERTICAL STRIPS, THEN OVER ONE, UNDER TWO, REPEATING. As you weave these strips in using the correct sequence, you will see that you are creating diamond shapes out of the layer 2 strips. The diamonds will touch tip-to-toe, all in a 30° row.

6. Repeat Steps 1–5 until the second layer covers the entire first layer, making sure to fill in the corners. *fig E*

When you reach the corners, you may need to temporarily unpin some first layer strips before you can weave the rest of the way.

THREADING THE WEFTY

The WEFTY is designed with a tapered end to glide through several layers of folded fabric strips. To do this effectively, thread your WEFTY for the third layer differently than you have been by holding it with the numbers facing up.

THIRD LAYER

The third layer is a similar order to the second layer but starts at a different fabric number and is woven from the left side at a 30° angle. The first strip that begins on the bottom right corner is Fabric 6. **THE SEQUENCE OF HOW IT IS WOVEN INTO THE FIRST LAYER IS UNDER ONE, OVER TWO. MOVING UP, FABRIC 7 WILL BEGIN OVER ONE, UNDER ONE, OVER TWO, UNDER ONE. FABRIC 8 WILL BEGIN OVER TWO, UNDER ONE.** *fig F*

You may find the shape that the cluster of two second layer strips and one first layer strip makes: to some people it looks like a backwards letter "z" or a lightning bolt.

Use That Purple Thang by inserting it under the second layer strip on the right side of the shape and letting it act as a ramp for the WEFTY which is entering the left side of the shape.

Your WEFTY and That Purple Thang should meet underneath the center of the first layer strip. You can't see them meet up under there, but you will hear a "click" if they find one another. If you feel resistance or notice the WEFTY get caught in the strip folds, do not push through. Instead, back up, adjust, and try again.

Follow the photo to find the third layer color placement. *fig G*

securing the weave

1. Press the panel gently with an iron on the steam setting, being mindful to avoid the pins. This will begin activating the fusible in the interfacing and securing woven strips.

2. Place painter's tape around the perimeter and finger-press firmly in place.

3. Remove the pins by holding the strip down with one finger while pulling out the pin. Make sure all pins are removed!

4. Press gently around the edges of the woven panel with the iron on the steam setting.

5. Flip the panel over and steam press it through the interfacing until it is set everywhere. Move the panel to your sewing machine and baste stitch using the edge of the tape as a guide. Remove tape when finished.

6. Using the frame insert as a template, lightly mark a cutting line on the interfacing side of the woven panel. Baste again ⅛″ inside the marked line. Trim on the marked line. *fig H*

7. Affix the interfacing side of the weave to the frame insert and place in frame to enjoy on your wall for years to come.

G

H

WEAVE INTO DENIM JACKET BACK PANEL

FINISHED SIZE:
Varies based on what you weave into

Give something denim from your wardrobe a new lease on life by weaving into a part of it! You can weave the front panels of a jacket or some knee patches in jeans. For this project, we will be weaving into the back of a jacket. Once you get the basics down you can upcycle any denim or corduroy piece from your closet with some weaving!

TARA CURTIS OF WEFTY WOVE INTO THIS GAP DENIM JACKET USING GOLDEN HOUR PRINTS BY RUBY STAR SOCIETY FOR MODA FABRICS.

materials and supplies

MATERIALS

Fabric for weaving (for a large jean jacket): ⅝ yard total

Lightweight woven interfacing: ⅓ yard

SUPPLIES

Basic tools (page 7)

½″ Sasher or bias tape maker

½″ thick foam core board

½″ WEFTY Needle

Self-healing cutting mat (*optional*)

Rotary cutter or fabric shears

cutting

WEAVE FABRICS

• Cut 18 strips 1″ × width of fabric. I cut 3 strips each from 6 different fabrics.

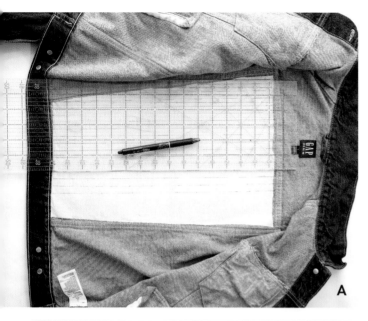

preparing the strips

1. Using the instructions on Preparing the Strips (page 14), press 18 beautiful strips ready for weaving.

2. Divide these strips evenly for layers 2 and 3. Remember that the denim is your first layer and we'll call it Fabric 1.

preparing the denim

1. Reinforce the back of the denim by fusing it with the lightweight woven fusible interfacing. This will mitigate some of the denim fraying.

2. Using a clear ruler and fabric marking pen or chalk, draw vertical lines every ½″ across the wrong side (or back) of the denim you will be weaving into. *fig A*

3. Lay your jacket on top of your cutting mat with the marked lines showing. Using the ruler and rotary cutter, or a pair of scissors, cut along each drawn line. *fig B*

4. Lay your foam core board on top of the jacket. Wrap the sides around the edges of the board, and then gripping the wrapped sides, flip the board and jacket around so that the front of your denim is facing you.

5. Pin in place. *fig C*

weaving

FIRST LAYER

The triaxial weave we will be doing is similar to tumbling blocks, with the denim acting as our first layer.

THREADING THE WEFTY

Thread your WEFTY by holding it with the WEFTY logo facing up and inserting a strip into the eye of the needle, folding it back with raw edges together.

SECOND LAYER

The second layer of the woven panel is made up of Fabric 2, woven into the first denim layer, from right to left, at 30°. Use That Purple Thang to hold up Fabric 1 (denim) strips for the WEFTY as you weave.

In order to ensure the strip is at 30°, you can use a piece of painter's tape laid down at the correct angle, or a ruler. *fig D*

C

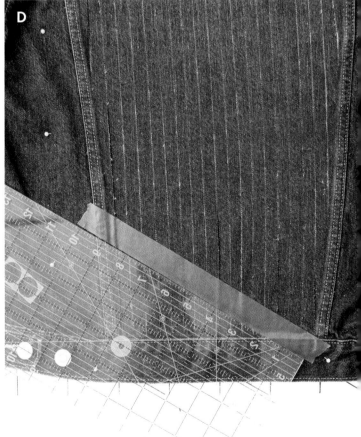

D

1. WORKING FROM THE RIGHT, WEAVE YOUR FIRST FABRIC 2 STRIP INTO THE FIRST DENIM LAYER USING THE FOLLOWING SEQUENCE: OVER ONE, UNDER TWO, REPEATING.

2. Tuck one end of the strip under the edge of your denim and pin. Trim the excess of the other end of the strip just outside of the other edge of your denim, then tuck and pin. You want the ends of your fabric strips to be hidden underneath your denim. *fig E*

3. Weave another Fabric 2 strip into the first layer right above the strip you just wove in.

WEAVE UNDER THE FIRST VERTICAL FABRIC 1 STRIP, THEN USE THE SEQUENCE OVER ONE, UNDER TWO, REPEATING.

4. Gripping each side of the strip, gently shimmy the strip back and forth to wiggle it closer to the first strip. Starting in the center of the strip, use That Purple Thang to finish pushing the strip close to the first, making sure there are no gaps between them, then pin.

5. **THE THIRD FABRIC 2 STRIP IS WOVEN IN UNDER THE FIRST TWO VERTICAL STRIPS, THEN OVER ONE, UNDER TWO, REPEATING.** As you weave these strips in using the correct sequence, you will see that you are creating diamond shapes out of the Fabric 2 strips. The diamonds will touch tip-to-toe, all in a 30° row.

6. Repeat Steps 1–5 until the second layer covers the entire first layer, making sure to fill in the corners.

E

THREADING THE WEFTY

The WEFTY is designed with a tapered end to glide through several layers of folded fabric strips. To do this effectively, thread your WEFTY for the third layer differently than you have been by holding it with the numbers facing up.

THIRD LAYER

The third layer is made up of Fabric 3 strips and is woven in from the left side at a 30° angle. **THE SEQUENCE OF HOW IT IS WOVEN INTO THE FIRST LAYER IS UNDER ONE, OVER TWO.**

You may find the shape that the cluster of two second layer strips and one first layer strip makes: to some people it looks like a backwards letter "z" or a lightning bolt.

Insert the WEFTY Needle from underneath the left side of the denim to start weaving. *fig F*

Use That Purple Thang by inserting it under the second layer strip on the right side of the shape and letting it act as a ramp for the WEFTY which is entering the left side of the shape.

Your WEFTY and That Purple Thang should meet underneath the center of the first layer strip. You can't see them meet up under there, but you will hear a "click" if they find one another. If you feel resistance or notice the WEFTY get caught in the strip folds, do not push through. Instead, back up, adjust, and try again.

When you reach the corners, you will need to temporarily unpin some strips to get the third layer strip woven in.

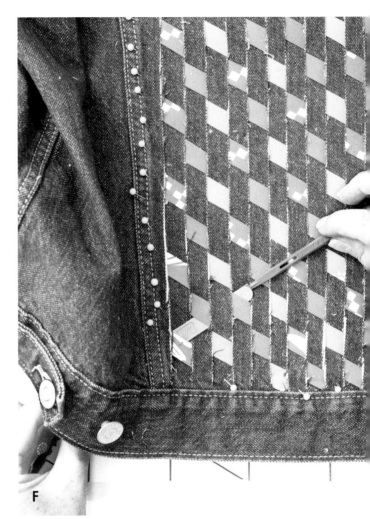

F

securing the weave

1. Secure the top of the weave using painter's tape. *fig G*

2. Remove the pins by holding the strip down with one finger while pulling out the pin. Make sure all pins are removed!

3. Lay another foam core board or piece of wood over the weave and board you were weaving on. Flip all three over and lift off the board you were weaving on. Secure the back of the weave with painter's tape. *fig H*

4. Flip again so you can see the top of the weave. Remove enough painter's tape to begin adding some stitching to permanently secure the weave. I sewed vertical lines every ½″ and around all four sides. *fig I*

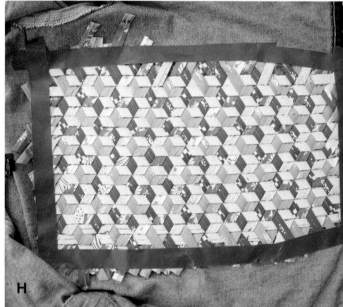

finishing

An optional finishing step is covering the raw edges by hand sewing leftover weaving strips over the four edges. *figs J–K*

You're going to really enjoy showing off your backside in this jacket!

J

K

WOVEN STARS MESSENGER BAG

FINISHED SIZE:
13″ × 12″ × 3″

Pick two contrasting fabrics to create six pointed stars. Make your panel into this cute messenger bag!

THIS WOVEN STARS MESSENGER BY TARA CURTIS/ WEFTY PAIRS TWO WOVEN CONTRASTING COLORS OF SPARK WITH A PRINT FROM STAY GOLD, ALL BY RUBY STAR SOCIETY FOR MODA FABRICS.

materials and supplies

MATERIALS

Fabrics for weaving: 1⅛ yard for stars (turquoise), ⅝ yard for outline (navy)

Lightweight woven interfacing: 1½ yards

Fabric for exterior and strap: 1⅛ yard. *Use a heavier substrate such as canvas or denim. If you choose a lighter substrate, reinforce the panel and gusset pieces with additional lightweight interfacing and reinforce the handle strap fabric with medium interfacing.*

Fabric for lining and pockets: ⅞ yard

Bias binding: 1 fat quarter (18″ × 22″)

Foam stabilizer: 20″ × 58″

SUPPLIES

Basic tools (page 7)

1″ Sasher or bias tape maker

½″ thick foam core board

1″ WEFTY Needle

(2) magnetic snap sets

(2) 1½″ D-rings

8″ zipper (*optional*)

1½″ slider

cutting

WEAVE FABRICS

- Cut 18 strips 2″ × width of fabric from stars fabric.

- Cut 9 strips 2″ × width of fabric from outline fabric.

LIGHTWEIGHT WOVEN INTERFACING

- Cut 1 rectangle 16″ × 20″ for woven panel.

- Cut 1 strip 3½″ × 33″ for gusset.

- Cut 1 rectangle 13″ × 19″ for back/flap lining.

- Cut 1 rectangle 12″ × 13″ for interior slip pocket.

- Cut 1 rectangle 10½″ × 13″ for lining front panel.

EXTERIOR FABRIC

- Cut 1 rectangle 10½″ × 13″ for front panel.

- Cut 2 strips 6″ × width of fabric for straps.

- Cut 1 strip 3½″ × 33″ for gusset.

LINING FABRIC

- Cut 1 rectangle 13″ × 19″ for back/flap piece.

- Cut 1 rectangle 10½″ × 13″ for front panel.

- Cut 1 rectangle 12″ × 13″ for interior slip pocket.

- Cut 2 rectangles 7″ × 9″ for zipper pocket lining (optional).

- Cut 1 strip 3½″ × 33″ for gusset.

BIAS BINDING FABRIC

- From the fat quarter, cut an 18″ square. Cut bias binding strips by folding the square diagonally and pressing. Align the 1¼″ mark of your ruler along the fold and cut. This strip when unfolded will measure 2½″. Cut 2½″ strips along the diagonal lines of the entire piece.

FOAM INTERFACING

- Cut 1 rectangle 13″ × 19″ for back/flap piece.

- Cut 1 rectangle 10½″ × 13″ for front panel.

- Cut 1 strip 3½″ × 33″ for gusset.

preparing the strips

1. Using the instructions on Preparing the Strips (page 14), press 18 beautiful strips ready for weaving from the stars fabric and 9 from the outline fabric.

2. Cut 4 stars strips and 3 outline strips in half so they measure approximately 20″ long.

preparing the board

1. Draw a 16″ × 20″ rectangle on your foam core board. Using the 30° mark on your ruler lined up along a 16″ edge of the rectangle, draw 2 angled lines in a bright color as shown in red and green in the illustration at right. Or, you can re-use the grid from another triaxial project. *fig A*

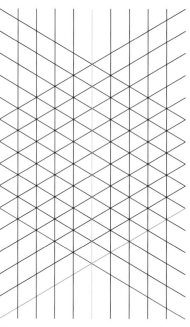

A

2. Use a dark color for the rest of your markings. Carefully position the ruler so that it rests vertically across the point where the two 30° lines intersect and draw a line down the center of the board.

3. Continue adding lines along the vertical and diagonals in 1″ intervals to create an isometric grid.

4. Lay the lightweight woven fusible interfacing *adhesive side up* over the grid. You should feel the bumpy side on the top! This is important because you don't want the bumpy side sticking to the board later. You want it to stick to the underside of the *woven strips* later. Pin each corner, inserting the pins at an extreme angle.

weaving

The stars weave pattern you are creating is made up of 3 layers. The first layer of your woven panel is made up of both stars and outline fabrics.

FIRST LAYER

Use the strips you cut in half to approximately 20″. Start with one outline fabric strip, then two star strips, one outline strip, two star strips, repeating, until you have 5 outline strips and 8 star strips on the board. You will have a total of 13 strips on the board, starting and stopping with an outline strip.

THREADING THE WEFTY

Thread your WEFTY by holding it with the WEFTY logo facing up and inserting a strip into the eye of the needle, folding it back with raw edges together.

SECOND LAYER

The second layer of the woven panel is woven into the first layer at 30°.

1. Thread your WEFTY with an outline strip and weave it into one side of the first layer strips, starting from the left side. Lift the first layer strips for the WEFTY using the Purple Thang.

THE WEAVING SEQUENCE YOU WILL FOLLOW IS OVER ONE, UNDER TWO, REPEATING. Ensure that the strip is straight along the 30° line by peeking between the first layer strips to see the gridlines.

2. Pull taut, and then pin at the very end of the strip close to the outside edge of the first layer. Pin just outside of the first layer on the other side of the Fabric 2 strip and pin. Trim this strip just next to the pin.

3. Weave a star fabric strip right next to the strip you just wove in.

START BY WEAVING UNDER THE FIRST STRIP, THEN USE THE FOLLOWING SEQUENCE: OVER ONE UNDER TWO, REPEATING.

4. Gripping each side of the strip, gently shimmy the strip back and forth to wiggle it closer to the first strip. Use a ruler and your Purple Thang to finish pushing the strip close to the first, making sure there are no gaps between them, then pin.

5. THE NEXT STRIP WILL BE ANOTHER STAR FABRIC STRIP. THE SEQUENCE IS: UNDER TWO, OVER ONE, REPEATING.

6. Repeat Steps 1–5 until you've woven into the entire first layer. The result will look awfully boring and monochromatic! Practice patience—the next layer will bring out the stars! *fig B*

THIRD LAYER

The third layer is woven in from the left side at a 30° angle.

Hold your WEFTY Needle with the number side facing up and thread it with a star fabric strip.

1. WEAVE THE STAR STRIP STARTING UNDER A SECOND LAYER OUTLINE STRIP, OVER TWO FIRST LAYER STRIPS, UNDER A SECOND LAYER STAR STRIP, AND OUT UNDER A SECOND LAYER OUTLINE STRIP. *fig C*

Pin and trim.

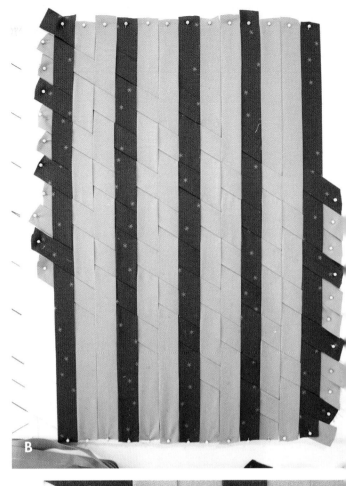

2. Weave the excess of the star strip right underneath the strip you just wove in. Start under the second layer outline strip, under the first layer strip, and out under the second layer star strip. *fig D*

D

3. The next two strips will be outine strips and will be woven above and below these two star strips. Once you have woven them in, you will the see the stars pop! Each outline strip is woven using this sequence: under a first layer outline strip, and out from under a second layer star strip, then over two first layer star strips and under the next second layer star strip, repeating.

4. Continue weaving in the third layer until the panel is complete. *fig E*

securing the weave

1. Press the panel gently with an iron on the steam setting, being mindful to avoid the pins. This will begin activating the fusible in the interfacing and securing woven strips.

2. Place painter's tape around the perimeter and finger-press firmly in place.

3. Remove the pins by holding the strip down with one finger while pulling out the pin. Make sure all pins are removed!

E

4. Carefully remove the panel from the board by gripping the sides of interfacing. Move the panel to your sewing machine and baste stitch around all 4 edges using a ⅛″ seam.

5. I quilted along the outer edges of where the stars strips are positioned, vertically, and diagonally, using matching thread. *fig F*

F

making bag components

LINING AND POCKETS

1. Fuse interfacing to lining pieces following manufacturer's instructions.

2. Fold the slip pocket piece in half wrong sides together, to 6″ × 13″, and press. Topstitch ⅛″ from the fold. Mark a vertical line down the center. Mark a line 1″ on either side of this line.

3. Lay slip pocket piece on the bottom of the front panel interior lining piece, making sure the fold is at the top. Match the raw edges of the pocket up to the raw edges of the lining. Beginning at the fold, sew the vertical center line you marked in Step 5, making sure to backstitch. This creates two slide pockets. Sew over the marked lines on either side of this line for pen pockets. Baste bottom and two sides of pocket to lining piece.

4. Baste foam stabilizer between the lining and exterior fabrics for the back/flap and gusset. Use a scant ¼″ seam allowance and pull the fabrics taut to prevent wrinkles. See Assembling the Bag (page 112) for instructions on rounding the corners of this piece. *fig G*

G

STRAPS

1. Join the handle strips right sides together at a 45° angle. Trim the excess at ¼″. See Binding, step 1, (page 10). Press seam open and trim dog ear triangles.

2. Lay the strip wrong side up, fold in half lengthwise, and press. Open strip and press edges to the middle fold. Fold down the middle lengthwise again and press. The strap will measure 1½″ wide. Topstitch ⅛″ from each edge down both sides of the strap.

3. Cut two 5″ pieces from the strap. Zigzag stitch along the raw edges of all strap pieces. Set the long strap aside for now.

4. Using a 5″ strap and D-ring, insert one end of the strap into the ring and fold it a third of the way up the strap. Fold the other third over this and clip or pin to hold. Repeat for the other strap and ring.

5. Sew the folded edges of these pieces to the two ends of the basted gusset piece by centering them 2½″ down from either end, ring towards gusset end. For best security, sew a square with an "X" inside of it. Go slow, as you are sewing through several layers, and avoid sewing through the hardware.

ADDING SNAPS

The magnetic snaps are placed in the top front panel corners and in the top corners of the back/flap piece.

1. On the exterior fabric of the front panel, mark the spot 2″ in from the sides and 4½″ down from the top.

2. On the lining fabric of the back/flap, mark the spot 2″ in from the sides, 1½″ down from the top.

3. Install the magnetic snaps where you marked, to the interior of the flaps and exterior of the front panel.

PREPARE THE BINDING

3. Follow the general instructions at Binding, Steps 1 and 2 (page 10) to join your binding strips and press.

4. Attach binding to the short ends of gusset first. Align raw edges of bias binding to the raw edge of the exterior side of the gusset. Sew, using a ¼″ seam. Then wrap the fold around to the lining side and hand sew it down. Trim the excess. Repeat for the other end of the gusset.

5. Repeat this process to attach binding to the top edge of the front panel.

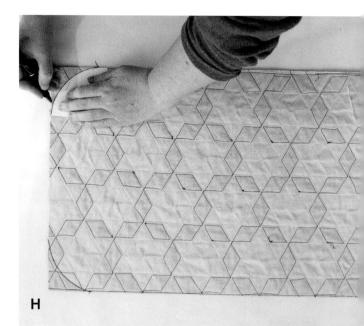

H

assembling the bag

1. Using pattern piece (page 142), mark the round bottom corners of front panel and all four corners of the back/flap piece. For added security, you can sew along the drawn line and then cut just outside it. *figs H–I*

2. Construct the body of the bag by laying the front panel piece lining side up. Lining sides together, align the gusset ends with the top corners of the front panel piece. Snip the fabric within the seam allowance around the curved parts of the gussets while you pin or clip the rest along the body piece. This will help the gusset ease in and reduce puckering.

3. Sew around the raw edges of the front panel and gusset using a ¼″ seam allowance.

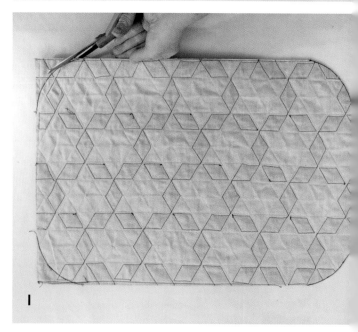

I

4. Line up the bottom of the free edge of the gusset and the bottom of the back/flap piece, wrong sides together. Pin the gusset to the back/flap piece in the same manner as the previous step. *fig J*

Ensure that the bag looks good and the gusset fits. Adjust if necessary.

5. Sew around the raw edges of the gusset using a ¼″ seam allowance.

6. Beginning at the bottom of back/flap and leaving a tail unsewn of 5″, attach bias binding. Leave room to join the bias ends. See Binding, steps 8 and 9 (page 11). *fig K*

7. Attach bias binding to front of bag, taking care to wrap the binding ends around the corners. Press binding around toward the gusset and hand stitch the folded edge to finish your bag.

J

K

GRADIENT DIAMONDS WOVEN CLUTCH

FINISHED SIZE:
9½″ × 6½″ closed
9½″ × 12″ open

The Gradient Diamonds weave is similar to Tumbling Blocks, however the first two layers are the same fabric. The third layer contrasts to showcase rows of diamonds. With color and texture galore, this little clutch will be as much fun to show off as it is to hold!

THIS WOVEN CLUTCH BY TARA CURTIS/WEFTY CONTRASTS FIVE WARM, ANALOGOUS COLORS WITH A COOL. ALL FABRICS ARE SPECKLED BY RUBY STAR SOCIETY FOR MODA FABRICS.

materials and supplies

MATERIALS

Fabrics for weaving: ⅙ yard each of 4 gradients and ⅔ yard for background

Fabric for lining and pockets: ½ yard

Fabric for bias binding and gussets: ¼ yard

Lightweight woven interfacing: 1 yard

Fusible fleece or foam stabilizer: ⅓ yard (optional for adding structure to the exterior body of bag)

SUPPLIES

Basic tools (page 7)

1″ **Sasher or bias tape maker**

½″ **thick foam core board**

1″ **WEFTY Needle**

¾″ **magnetic snap set**

Seam ripper

cutting

WEAVE FABRICS

NOTE *Do not trim selvages.*

- Cut 3 strips 1″ × width of fabric from each of the darkest 3 gradient fabrics (dark to light, Fabric 1, Fabric 2, and Fabric 3).

- Cut 4 strips 1″ × width of fabric from the lightest gradient fabric, Fabric 4.

- Cut 22 strips 1″ × width of fabric from the background fabric.

LINING FABRIC

- Cut 2 rectangles 7″ × 15″ for gussets.

- Cut 1 rectangle 9½″ × 19″ for the lining.

- Cut 1 rectangle 9½″ × 22″ for the pockets.

BIAS BINDING AND GUSSETS FABRIC

- Cut 1 square 9″ × 9″. Cut bias binding strips by folding the 9″ square diagonally and pressing. Align the 1¼″ mark of your ruler along the fold and cut. This strip when unfolded will measure 2½″. Cut 2½″ strips along the diagonal lines of the entire piece.

- Cut 2 rectangles 7″ × 15″ for gussets.

LIGHTWEIGHT WOVEN INTERFACING

- Cut 1 rectangle 12″ × 20″ for the weave.

- Cut 1 rectangle 9½″ × 19″ for the lining.

- Cut 1 rectangle 9½″ × 22″ for the pockets.

- Cut 2 rectangles 7″ × 15″ for gussets.

FUSIBLE FLEECE OR FOAM INTERFACING

- Cut 1 rectangle 9½″ × 19″.

preparing the strips

1. Using the instructions on Preparing the Strips (page 14), press 3 beautiful strips ready for weaving from each color of Fabric 1, 2, and 3. Do the same with 4 strips of Fabric 4.

2. Repeat the process with the background fabric until you have 22 strips. Fold 9 background strips in half and cut to get 18 strips that each measure approximately 20˝ long. Keep your strips organized.

preparing the board

1. Draw a 12˝ × 20˝ rectangle on your foam core board. Using the 30° mark on your ruler lined up against a 12˝ edge of the rectangle, draw 2 angled lines in a bright color as shown in red and green in the illustration at right. *fig A*

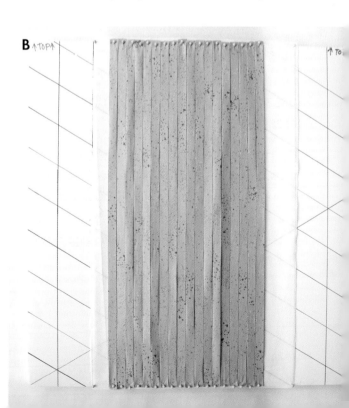

A

2. Use a dark color for the rest of your markings. Carefully position the ruler so that it rests vertically across the point where the two 30° lines intersect and draw a line down the center of the board.

3. Continue adding lines along the vertical and diagonal in 1˝ intervals to create an isometric grid.

4. Lay the lightweight woven fusible interfacing *adhesive side up* over the grid. You should feel the bumpy side on the top! This is important because you don't want the bumpy side sticking to the board later. You want it to stick to the underside of the *woven strips* later. Pin each corner, inserting the pins at an extreme angle.

weaving

The triaxial weave pattern you are creating is made up of 3 layers. The first 2 layers of your woven panel are made up of the background fabric.

FIRST LAYER

Lay a background fabric strip down on the board, raw edges facing down. You should be able to see the gridlines through the interfacing. Use them as guides to keep the strip straight. Pull the strip taut and pin, ensuring you have at least 20˝ of space between the pins. Repeat until you have pinned 18 strips. *fig B*

B

THREADING THE WEFTY

Thread your WEFTY by holding it with the WEFTY lettering up and inserting a strip into the eye of the needle, folding it back with raw edges together.

SECOND LAYER

The second layer of the woven panel is woven into the first layer, from right to left, at a 30° angle. Use That Purple Thang to hold up first layer strips for the WEFTY as you weave.

1. WORKING FROM THE RIGHT, WEAVE YOUR SECOND LAYER STRIP INTO THE FIRST LAYER USING THE FOLLOWING SEQUENCE: OVER ONE, UNDER TWO, REPEATING.

Ensure the strip follows along your brightly colored 30° line by peeking between the first layer strips to see the gridlines.

2. Pull taut, and then pin at the very end of the strip close to the outside right edge of the first layer. Then pin just outside of the first layer on the other left side. Trim the excess of this strip just next to the pin.

3. Weave the excess of the strip into the first layer right above the strip you just wove in.

WEAVE UNDER THE FIRST VERTICAL STRIP, THEN USE THE SEQUENCE OVER ONE, UNDER TWO, REPEATING.

4. Gripping each side of the strip, gently shimmy the strip back and forth to wiggle it closer to the first strip. Starting in the center of the strip, use That Purple Thang to finish pushing the strip close to the first, making sure there are no gaps between them, then pin.

5. THE THIRD SECOND-LAYER STRIP IS WOVEN IN UNDER THE FIRST TWO VERTICAL STRIPS, THEN OVER ONE, UNDER TWO, REPEATING. As you weave these strips in using the correct sequence, you will see that you are creating diamond shapes out of the second layer strips. The diamonds will touch tip-to-toe, all in a 30° row.

6. Repeat Steps 1–5 until the second layer covers the entire first layer, making sure to fill in the corners.

When you reach the corners, you may need to temporarily unpin some first layer strips before you can weave the rest of the way. *fig C*

THREADING THE WEFTY

The WEFTY is designed with a tapered end to glide through several layers of folded fabric strips. To do this effectively, thread your WEFTY for the third layer differently than you have been by holding it with the numbers facing up.

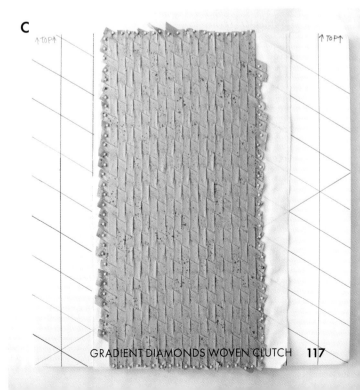

C

third layer

The third layer is made up of Fabric 1, 2, 3, and 4 strips and is woven in from the left side at a 30° angle. The sequence of how it is woven into the first layer is under one, over two.

You may find the shape that the cluster of two second layer strips and one first layer strip makes: to some people it looks like a backwards letter "z" or a lightning bolt.

Use That Purple Thang by inserting it under the second layer strip on the right side of the shape and letting it act as a ramp for the WEFTY which is entering the left side of the shape.

Your WEFTY and That Purple Thang should meet underneath the center of the first layer strip. You can't see them meet up under there, but you will hear a "click" if they find one another. If you feel resistance or notice the WEFTY get caught in the strip folds, do not push through. Instead, back up, adjust, and try again.

Follow the arrows in the illustration below to find the third layer placement. Fill in the top and bottom corners with Fabric 1.

1. Start weaving a fabric 1 strip from the bottom-most lefthand corner, and weave until that corner is filled. Flip the board around so that the corner you filled in is now at the top. Repeat what you just did for the corner now closest to you. *figs D–E*

2. Weaving from the corners with Fabric 2, fill in 4 rows on either side. *fig F*

3. Fill in 4 more rows with Fabric 3. *fig G*

4. Fill in the center with Fabric 4. *fig H*

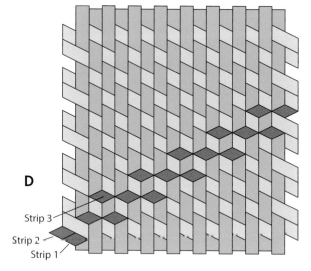

D

Strip 3

Strip 2

Strip 1

securing the weave

1. Press the panel gently with an iron on the steam setting, being mindful to avoid the pins. This will begin activating the fusible in the interfacing and securing woven strips.

2. Place painter's tape around the perimeter and finger-press firmly in place.

3. Remove the pins by holding the strip down with one finger while pulling out the pin. Make sure all pins are removed!

4. Carefully remove the panel from the board by gripping the sides of interfacing. Move the panel to your sewing machine and baste stitch around all 4 edges using a ⅛″ seam.

5. Round the corners of one end of the woven panel using the pattern piece (page 142).

6. Baste stitch just inside of this drawn line, then cut just outside of your stitches. This is the top of your bag exterior piece.

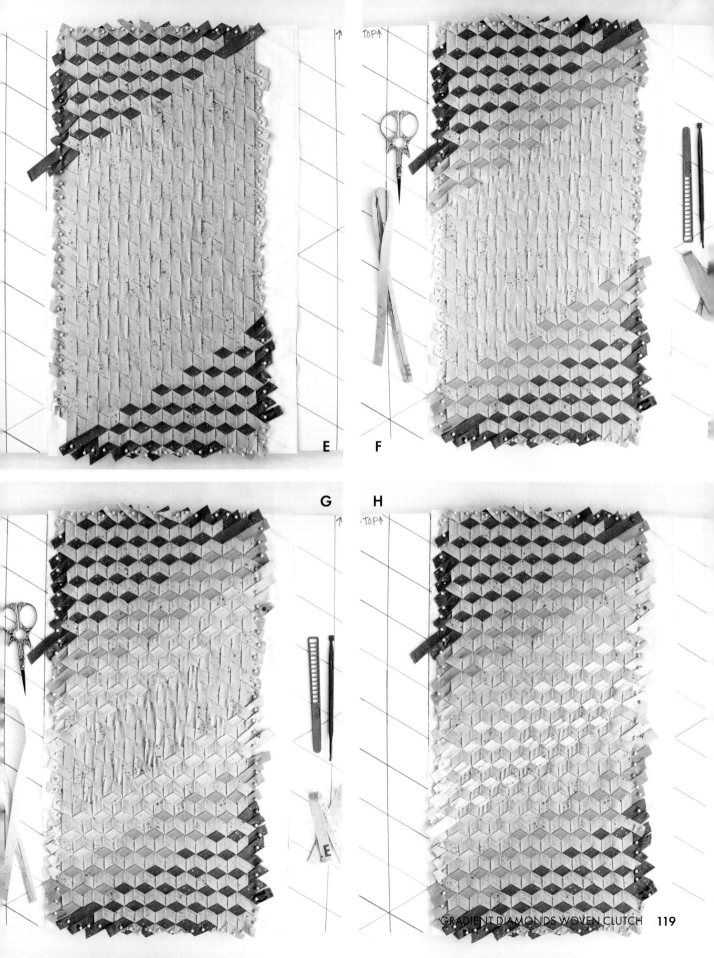

E

F

G

H

I

constructing the interior

LINING

Fuse the 9½″ × 19″ lightweight woven fusible interfacing to the wrong side of the 9½″ × 19″ lining piece. Round two corners on one end using the pattern piece (page 142). This is the top of your lining piece. *fig I*

POCKETS

1. Fold the 9½″ × 22″ piece of lining fabric in half, right sides together to make a piece 9½″ × 11″. Sew along the raw 9½″ edges using a generous ¼″ seam. Press open. Turn this piece right side out and press.

2. Topstitch along the folds on either end of the pocket piece. Mark the raw-edge side of the pocket piece at 5¼″ from one folded end.

3. To attach the pocket to the lining piece, line up one topstitched end of the pocket piece 1½″ from the bottom (not rounded end) of the lining piece. Pin in place.

4. Sew the pocket to the lining widthwise at the 5¼″ mark. Then add sewn lines ½″ on either side of this sewn line. *fig J*

OPTIONAL STRUCTURE

If you are using fusible fleece or foam interfacing to add structure to your bag, use the corner pattern piece to round two corners on one end of your 9½″ × 19″ rectangle.

J

adding magnetic snap

You must add the magnetic snap before sewing the bag together. You may reinforce the areas where the magnetic snap will be installed with a piece of fusible interfacing if you like.

The female part of the magnetic snap will be installed on the bottom of the woven side, and the male part will be installed on the top of the lining side.

1. Mark the wrong side of the bottom of your woven panel, 3¼″ from the bottom edge, 4¾″ from either side.

2. Using the washer, mark lines where the prongs will go through. Use a seam ripper to cut the lines all the way through the woven strips. Make sure not to make your hole bigger than the line you drew.

3. Insert the female side of the magnetic snap through the washer, and fold the prongs to either side to attach it.

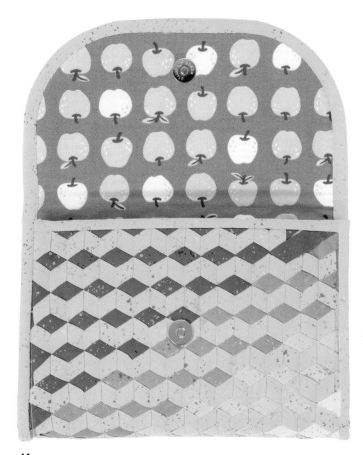

K

4. The male side will be attached 1½″ from the top of the lining, 4¾″ from either side. Install it the same way you installed the female side. *fig K*

bag body piece

1. Layer the interior, the fusible fleece or foam interfacing (if using), and the exterior pieces wrong sides together and baste all the way around using a narrow zigzag stitch.

2. Fuse the 7″ × 15″ piece of lightweight fusible interfacing to the 7″ × 15″ gusset fabric piece. Fold in half for a piece that measures 7″ × 7½″ and press. Topstitch along the fold. Cut the gussets out using the pattern piece (page 142). Make sure to line the top of the gussets along the topstitched fold before cutting. Baste stitch each gusset closed around the raw edges.

3. Measure 5¾″ from the top of the lining side of the bag body and mark the edge of either side. This is where the top of gussets will be attached when you are ready.

binding

MAKING THE BINDING

Join the ends of the binding strips right sides together (see Binding, page 10) and trim off the dog ears. Lay the strip face down on your ironing board, fold in half lengthwise and press to get 2½˝ single-fold bias binding.

ATTACHING THE BINDING

You need to first add binding to the bottom of the body piece before adding the gussets.

1. Align the raw edge of the binding to the raw edge of the bottom of the woven side of the body piece. Each end of the binding will be left open. The raw edges will all be hidden after we finish sewing the bag together.

2. Sew using a ¼˝ seam. Fold the binding around to the lining side of the bag body piece and press. Hand stitch the fold down to the lining side.

constructing the bag

ATTACHING THE GUSSETS

1. Align one side of the top of the gusset with the mark you made on the lining. Clip and/or pin. *fig L*

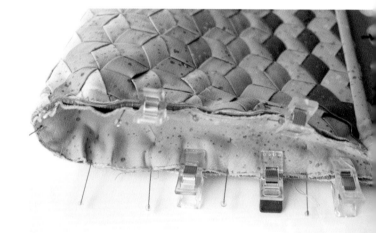

L

2. Fold the bag piece around and align the other side of the top of the gusset with the corner of the bottom of the bag. Clip and/or pin.

3. Snip the fabric within the seam allowance around the curved parts of the gussets while you pin the rest along the body piece. This will help the gusset ease in and reduce puckering.

4. Sew carefully and slowly. Take your time, as this is several layers for your machine to go through. *fig M*

5. Repeat Steps 1–4 to attach the other gusset.

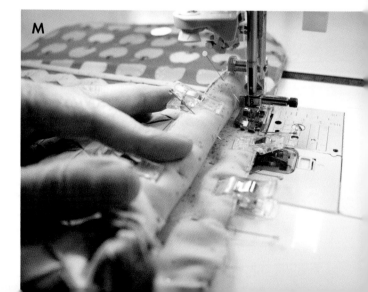

M

SEWING ON THE BINDING

1. Attach binding to the woven exterior of the bag starting at a front corner, leaving a 2˝ tail at the beginning. Sew carefully all the way around to the last front corner, leaving a 2˝ tail at the end. *fig N*

2. Fold the binding toward the gussets on the sides, and toward the lining side of the flap. Press and clip.

3. At the front corners, fold the end of the binding tail under until it is hidden underneath itself. Trimming at an angle helps this. *fig O*

4. Hand sew to finish the binding.

Your new clutch is ready for date night!

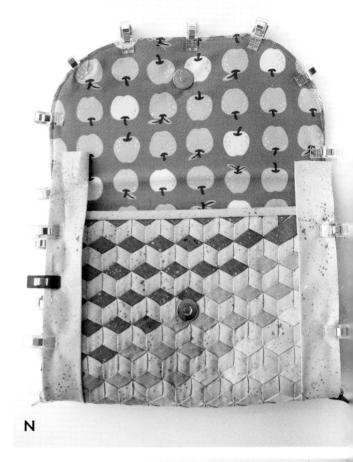

N

O

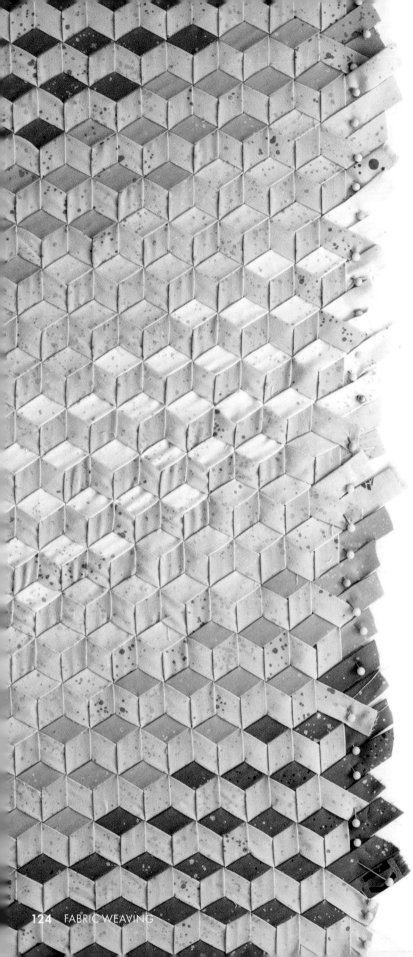

CONCLUSION

If you've made it to this portion of the book, you've at a minimum flipped through all the pages to see the awesome weaves and projects. Now you're probably reading this trying to decide which one to start with...and we get it. With biaxial weaving, we gave you six different weaves ranging from twill to tartan to houndstooth to an optical illusion of awesomeness. And then with triaxial weaves, you could choose the ubiquitous tumbling blocks or an abstract kaleidoscope or even slice up a jacket and weave into it. The added bonus of not knowing which weave to do is that you absolutely can't go wrong with any of them.

While we legit hate making this decision even harder for you, we also gave you 13 different projects that you can make. And guess what? You don't have to pair the projects with the weaves the way that we did. If you want to make a Tumbling Blocks Tote, have at it. A Silly Silo Clutch would be super fetch. Basically, this book is filled with an infinite number of possible creations.

The creativity doesn't have to stop there either. Minor tweaks to color, print, or order can have dramatic results, so once you've gotten the techniques and ideas down, you can use those skills to develop your own creations. We have provided a gallery of inspiration to help the creative juices get flowing. Happy weaving!

GALLERY

BLUE CATCH-ALL CADDY by Mister Domestic *top*
OMBRÉ WOVEN STARS by Tara Curtis/WEFTY *below*
WOVEN AMERICAN FLAG by Mister Domestic *right*

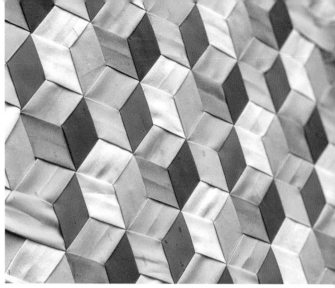

WOVEN STAR VARIATIONS TOTE by Tara Curtis/WEFTY *left*
TRANS PRIDE TUMBLING BLOCKS by Mister Domestic *top*
WOVEN STAR VARIATIONS by Tara Curtis/WEFTY *below*
SUNSET WEAVE by Mister Domestic *bottom*

WOVEN STAR VARIATIONS PILLOW
by Tara Curtis/WEFTY *below*

WOVEN DENIM TRENCH COAT by Mister Domestic *left*

WOVEN TABLE MAT by Tara Curtis/WEFTY *top*

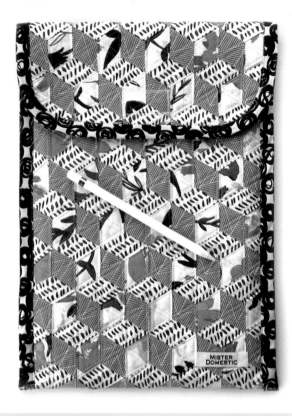

TUMBLING BLOCKS TABLET COVER by Mister Domestic *top*

WOVEN BASKET by Tara Curtis/WEFTY *bottom left*

KALEIDOSCOPIC ZIPPER CASE by Mister Domestic *right*

TRIAXIAL WEAVE by Tara Curtis/WEFTY *bottom right*

RAINBOW GRADIENT WOVEN PILLOW by Mister Domestic *bottom left*
TUMBLING BLOCKS by Tara Curtis/WEFTY *bottom right*
TWILL WEAVE IN LINEN by Mister Domestic *left*
BASKET WEAVE POUCH by Tara Curtis/WEFTY *top*

YARN-DYED WOVEN QUILT
by Mister Domestic *right*

KALEIDOSCOPE WALL ART
by Mister Domestic *top left*

**TUMBLING BLOCKS MESSENGER
BAG** by Mister Domestic *top right*

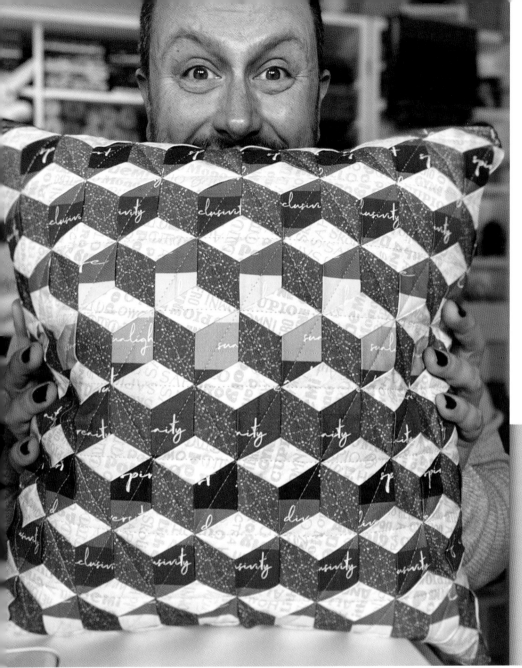

TUMBLING TRIPLETS LAPTOP CASE
by Tara Curtis/WEFTY *bottom left*

WOVEN DIAMONDS
by Tara Curtis/WEFTY *top*

RAINBOW STRIPED TUMBLING BLOCKS
by Mister Domestic *left*

WOVEN STAR VARIATIONS PILLOW
by Tara Curtis/WEFTY *bottom right*

LATTICE WEAVE CARRYING CASE
by Mister Domestic *top*

TUMBLING TRIPLETS TOTE
by Tara Curtis/WEFTY *top right*

WOVEN BASKET
by Mister Domestic *bottom*

WOVEN STAR VARIATIONS TOTE
by Tara Curtis/WEFTY *bottom right*

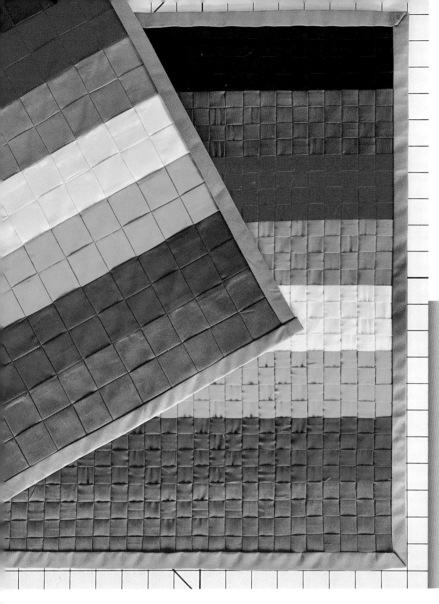

WOVEN IPAD CASE
by Tara Curtis/WEFTY bottom *left*

WOVEN PHILADELPHIA PRIDE FLAG
by Mister Domestic *left*

RAINBOW GRADIENT ART CASE
by Mister Domestic *above*

WOVEN IPAD CASE
by Tara Curtis/WEFTY *below*

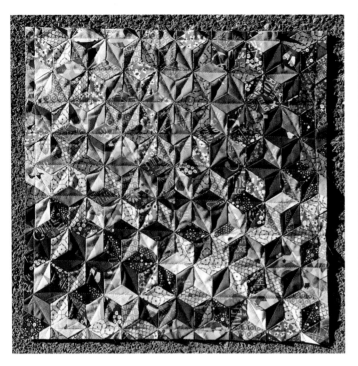

KALEIDOSCOPE PANEL
by Mister Domestic *top left, page 136*

KALEIDOSCOPE PANEL
by Mister Domestic *top right, page 136*

SELVAGE WOVEN POUCH
by Mandy Andy Designs *bottom, page 136*

OPTICAL ILLUSION WEAVE IN BLUES
by Mister Domestic *top right*

DUMPLINGS
by Tara Curtis/WEFTY *top left*

WEAVE YOUR JACKET
by Tara Curtis/WEFTY *left*

WOVEN DRESS BODICE
by Mister Domestic *top left, page 138*

LATTICE WEAVE PILLOWS
by Mister Domestic *top right, page 138*

CLARE ANNA BAG
by Tara Curtis/WEFTY *bottom, page 138*

BIAXIAL WEAVE
by Tara Curtis/WEFTY *below*

TAKE IT OR WEAVE IT BASKET
by Tara Curtis/WEFTY *left*

WOVEN GRADIENT DENIM QUILT
by Mister Domestic *bottom left*

DRESS BODICE
by Mister Domestic *right*

PURPLE OPTICAL ILLUSION PILLOW
by Mister Domestic *below*

WANDERER BAG
by Tara Curtis/WEFTY *bottom right*

WOVEN DIAMONDS MESSENGER BAG
by Tara Curtis/WEFTY *top left, page 141*

GRADIENT BLACK AND WHITE TUMBLING BLOCKS
by Mister Domestic *bottom left, page 141*

WOVEN STARS MESSENGER BAG
by Tara Curtis/WEFTY *right, page 141*

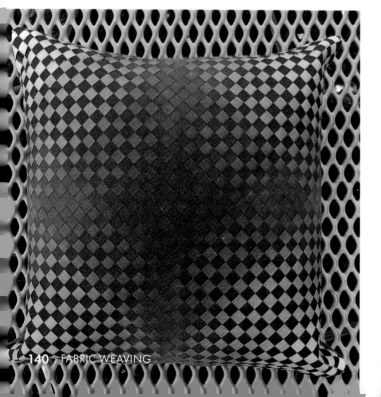

PATTERNS

Corner cutting template

Gusset
Cut 2 Gusset fabric.
Cut 2 Lining.
Cut 2 Interfacing.

ABOUT THE AUTHORS

Photo by Becca Blevins

Photo by Tara J. Curtis

MATHEW BOUDREAUX of Mister Domestic learned to sew as a kid, but the antiquated binary gender expectations of his parents got in the way, so he never really felt compelled to up his game. Shortly after his daughter Helena was born in 2013, his spouse bought him a couple sewing classes and then he took off like he was at the races. With his kid as his muse and inspiration, the quality and coolness of the stuff he made far exceeded anything that he thought he'd ever be able to create with his own hands. He's a fabric and pattern designer, sewing instructor, owner of his new online sewing school SEW U, inspirational speaker, consultant, and global influencer with his TikTok, YouTube, and Instagram each set to surpass over 100,000 this year. Hands down, his favorite thing about this journey is the truly inclusive Mister Domestic community that has been created by all its members.

TARA J. CURTIS is the inventor of the WEFTY Needle and works as a Digital Sales Specialist for Fiskars. While fabric weaving is her latest passion, upcycling secondhand clothing was her first. She loves being around fabric at all times and teaching folks how to weave. Tara's first career was advocating for the rights of rape victims. When someone experiences sexual assault, she starts by believing that person. Tara is a volunteer for the Social Justice Sewing Academy Remembrance project. If you'd like to make a quilt block to honor the life of a victim of authority-, community-, race-, or gender- and sexuality-based violence, please visit https://www.sjsacademy.org/remembrance-project